# 哥特建筑与雕塑装饰艺术

## 第2卷

曹峻川　甄影博　编译

[英]拉斐尔·布兰登　[英]约书翰·亚瑟·布兰登　绘

江苏凤凰科学技术出版社

# 前言

　　公元第二个千年开端不久，在诺曼王朝即将上台的前夕，英国的宗教建筑逐渐转变为后来在诺曼王朝中的独特样式，即我们现在所定义的"盎格鲁－诺曼"风格。这种风格最初由"忏悔者"爱德华介绍到英国，或者说是由卡纽特大帝，然后通过他们应用于自己领地中大量的教堂建设中。可以说，在诺曼王朝统治下迅速发展的教堂建筑，实际上在王朝来临之前就已经建立了相当完善的体系。盎格鲁－诺曼建筑的建筑师们竭尽所能，使其风格更加完美，从现存的一些建筑便可见一斑。尽管它们发展得如此之好，但建筑如同陆地上其他事物一样不能长久稳定。如我们所见，一种建筑形式或样式，一旦到达其成熟期就会或多或少被其他的建筑形式和样式所取代，这是一种潜在的规则。但是全然不顾这种潜在规则的影响，盎格鲁－诺曼王朝保留了大量古代传统建筑来宣称其永久的建立。低矮、笨重的比例，沉重且自承重的墙体，矩形的叠内拱，方形的柱顶板及柱础，以及严格说来肤浅的装饰——所有这些似乎都在诉说着一种从罗马退化到古罗马式的"宏伟"样式，而非真正从自身土壤中发展起来的伟大样式——中世纪的建筑准备用古代建筑的标准来衡量自身的力量。同时，在盎格鲁－诺曼时期，基督教迅速发展，大量的教堂建筑欠缺，仅有一些巴西利卡式的建筑——它们在不久后就由于异教徒的起源而被移除，尽管巴西利卡自身不是异教的。因此，在回顾盎格鲁－诺曼建筑的最终结束和及盎格鲁－哥特建筑建筑样式完全建立之间这段时期时，人们更多沉浸在过去的建筑样式中而不是追寻一种卓越的替代者。经历这段关于建筑样式的挣扎之后，新建筑样式的基本元素开始与旧的建筑特征相融合，哥特建筑逐渐获得了一种特定的形式，它更加明亮、深邃宽敞及高贵，相比于早期的英国样式，随即便展现出其优越性。

　　《哥特建筑手册》的作者曾评述到："这种样式如此优美，自身十分完美，或许可以想象在任何的历史时期或者地方能有与其并驾齐驱的建筑样式，或者建造其的匠人和说服各个教堂的智者没能做到这一点，未来的后代也不能看待能与其媲美的建筑。"

　　亨利三世（1207 年—1272 年）统治后期，哥特建筑在细部及组合方式上出现了一些新奇的方式。被所在墙体清晰分隔并由连续的披水石及披水饰结合在一起的尖顶窗，由大尺度并被竖框分隔成多个窗扇的窗户所取代；竖框的引入使得以丰富的几何形体布满窗户的花格窗饰也随之产生。线脚上粗的凸起与深的凹槽之间的交替让步于更加丰富和优美曲折的新组合方式；小柱子不再分散布置或者被捆绑成束，反而是更加坚实地连在一起；卷叶形花饰作为人们最喜爱的哥特装饰，更多地从自然树木及植物中吸取元素；不同于从一簇向上伸展的茎中伸出的波浪状三叶饰，几片叶子更趋向于一种环绕的形式，然后包围它们所附着的物体。更加丰富和更具差异性的装饰也从少数哥特建筑中往外蔓延，赋予哥特式建筑更加精巧的层次性。

如此哥特建筑逐渐从早期的英国样式进入到盛装哥特样式——也是最为人称赞的样式——与爱德华时代一起形成的完美的盎格鲁－哥特艺术。随着这种样式的发展，一些特征的差异性更加显著，同时，相对于早期对于几何形式精确性的追求，现在的哥特艺术更加倾向于优美的波浪状流动线条。

出现在早期英国哥特样式中的空间上垂直性的趋势与盎格鲁－诺曼时期罗马式的水平延展的样式产生强烈对比。在盛装哥特样式中，主要的结构线条形成一种锥形边缘，而非垂直或水平的。为了在这个基本规则下实现这一系列变化，盎格鲁－哥特样式第三个清晰的时期的特征便是由垂直线条以及与其正交的同等重要的线条所界定的。这个最新的华美样式，因其线条的突出地位而被命名为垂直哥特式，逐渐取代了盛装哥特样式。就像作为一个更加成熟的样式，盛装哥特由于其精美及和谐的丰富性逐渐替代早期英国的哥特样式一样。作为一种新的样式，垂直哥特式建筑暂时保留了部分之前样式的特征，并与其自身特殊的特征相结合：作为垂直哥特式的第一个时期，也想要达到盛装哥特式的壮美，但用太高的赞美来评价是很困难的。然而，随着都铎式建筑中平坦拱的出现，更加丰富多样和细致的嵌板及其它装饰也随之诞生，这也清晰表明了这个时代在建筑品味上的衰退。而建筑历史中的一次后退往往是致命的。

因此，中世纪的教堂建筑尽管在衰退，但整体还是很壮美，由扇形花格饰布满的拱顶频繁出现在最后几个伟大的作品中——之后，建筑史中长时间的衰退时代就来临了。

曹峻川　甄影博

# 目 录

第 $1$ 章

# 约克郡地区 教堂建筑

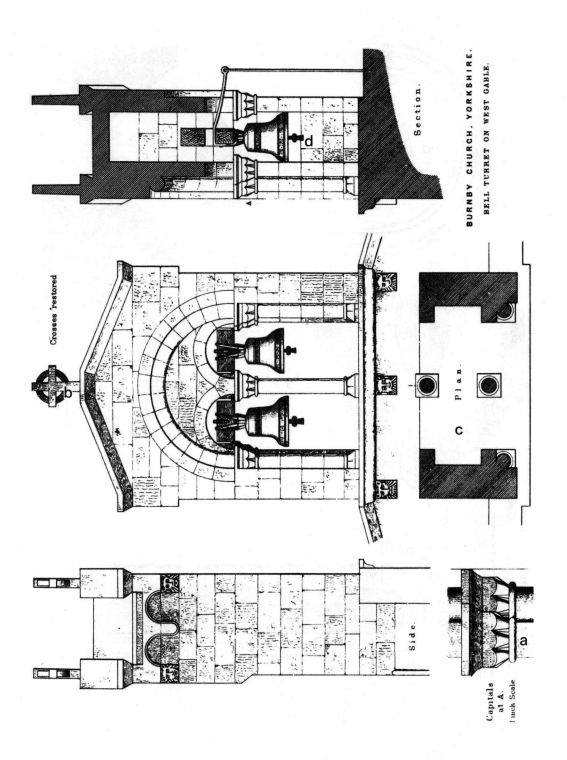

约克郡本拿比教堂：a.柱头　b.十字架复原　c.平面剖面　d.西侧山墙钟楼

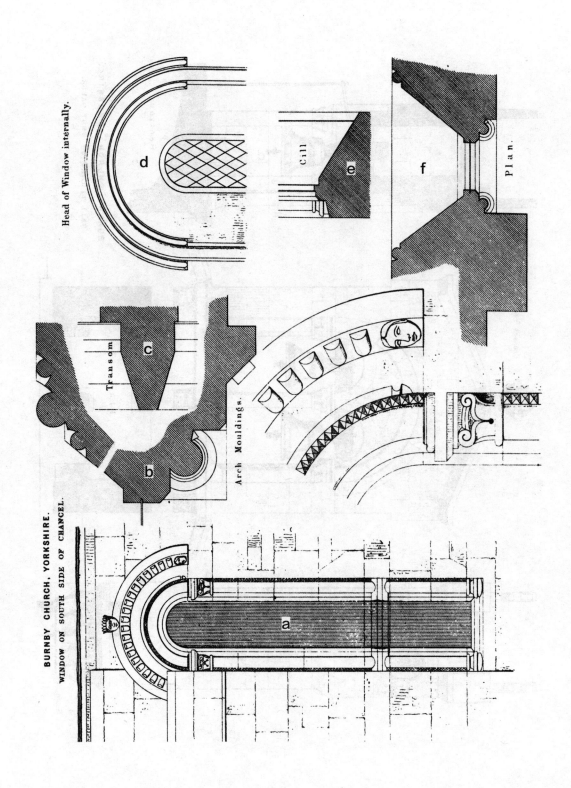

约克郡本拿比教堂：a. 圣坛南侧窗户  b. 拱门线脚  c. 横楣  d. 内部窗顶  e. 平面  f. 门槛

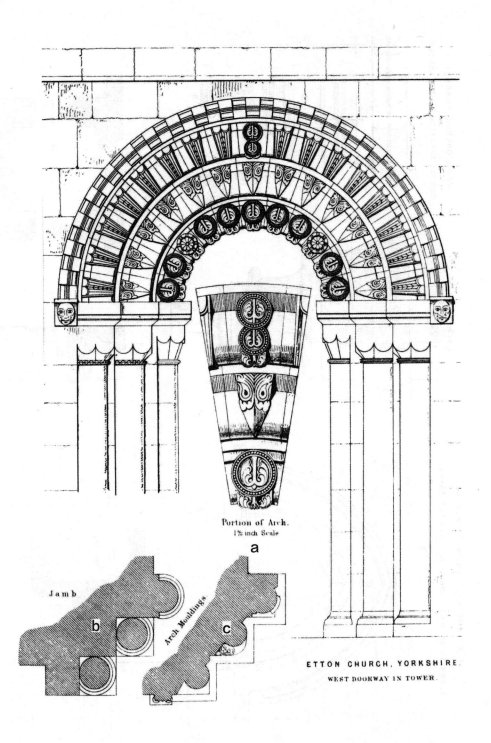

Portion of Arch.
1½ inch Scale

a

Jamb

b

Arch Mouldings

c

ETTON CHURCH, YORKSHIRE.
WEST DOORWAY IN TOWER.

约克郡艾通教堂塔楼西门：a. 拱门局部　b. 侧壁　c. 拱门线脚

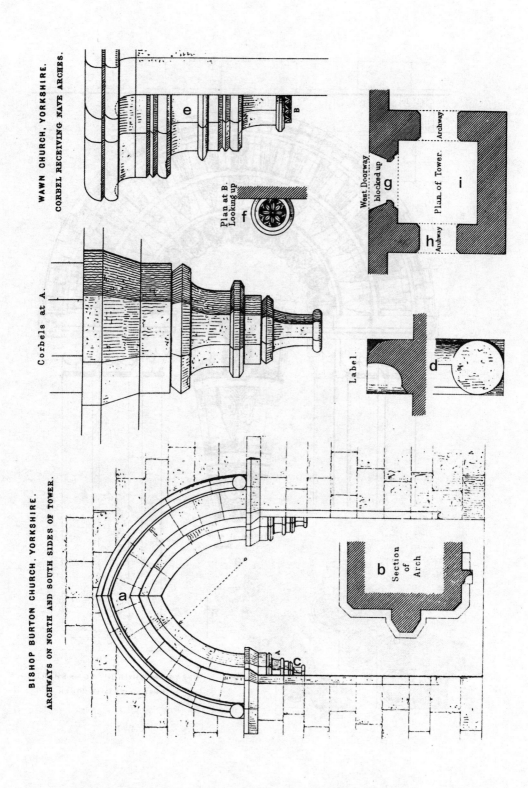

约克郡伯顿主教堂：a. 塔楼南北两侧拱道　b. 拱门剖面　c.A 处梁托　d. 披水石

约克郡万恩教堂：e. 承载中殿拱的梁托　f.B 处仰视平面　g. 被封堵的西门　h. 拱门　i. 塔楼平面

Label and Mullion.
1½ Inch Scale.

a

b

Label Termination. ¼ full size.

AISLE WINDOW, COTTINGHAM CHURCH, YORKSHIRE

c

Cill

d

Plan

约克郡科廷厄姆教堂：a. 披水石和竖框　b. 披水石端部　c. 窗台　d. 平面

WAWN CHURCH, YORKSHIRE,
WEST WINDOW OF NAVE.

Jamb and Mullion
of Window.
1inch Scale

约克郡万恩教堂：a. 中殿西侧窗户　b. 窗户侧壁及竖框

第 2 章

埃塞克斯郡地区
教堂建筑

TILTEY CHURCH, ESSEX.

NORTH WINDOW OF CHANCEL

埃塞克斯郡迪尔帝教堂：a. 圣坛北侧窗户　b. 披水石　c. 侧壁　d. 叶形饰的尖头　e. 窗台剖面　f. 平面

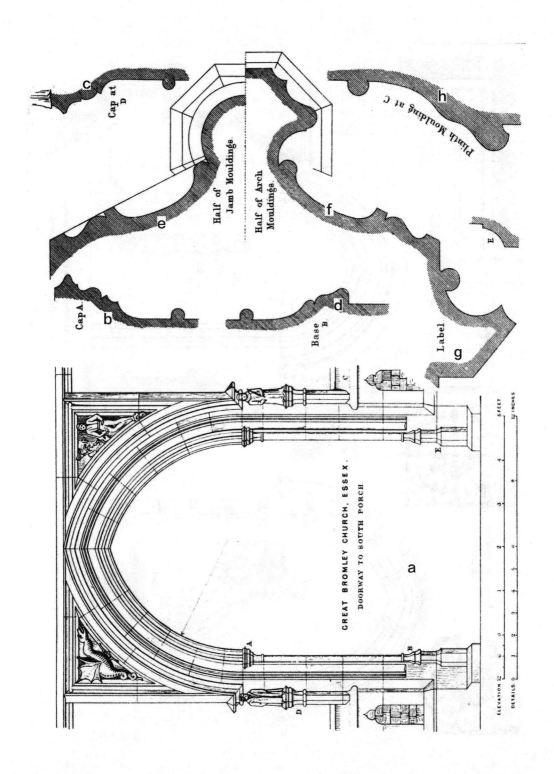

埃塞克斯布罗姆利大教堂：a. 通向南侧门廊大门　b.A 柱头　c.D 柱头　d.B 处柱础　e. 侧壁线脚一半　f. 拱线脚一半　g. 披水石　h.C 处底座线脚

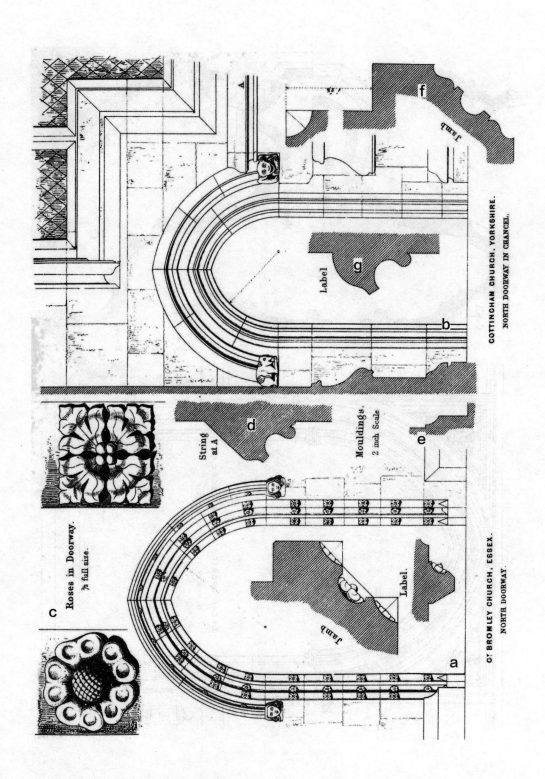

埃塞克斯布罗姆利大教堂:a.北门约克郡科廷厄姆教堂　b.礼拜堂北门　c.蔷薇装饰　d.A处的束带层　e.线脚　f.侧壁　g.披水石

LITTLE TOTHAM
SOUTH

CHURCH, ESSEX.
DOORWAY.

a

Arch Mouldings

b

Jamb

c

Detail of Arch.

d

埃塞克斯托特汉小教堂：a. 南侧大门　　b. 拱线脚　　c. 侧壁　　d. 拱细部

第 ③ 章

# 诺福克郡地区教堂建筑

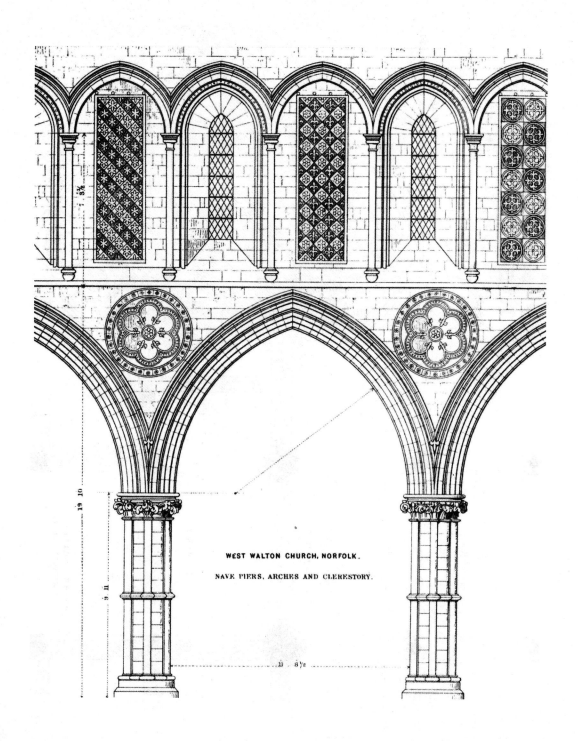

WEST WALTON CHURCH, NORFOLK.

NAVE PIERS, ARCHES AND CLERESTORY.

诺福克西沃尔顿教堂中殿集束柱、拱和侧高窗

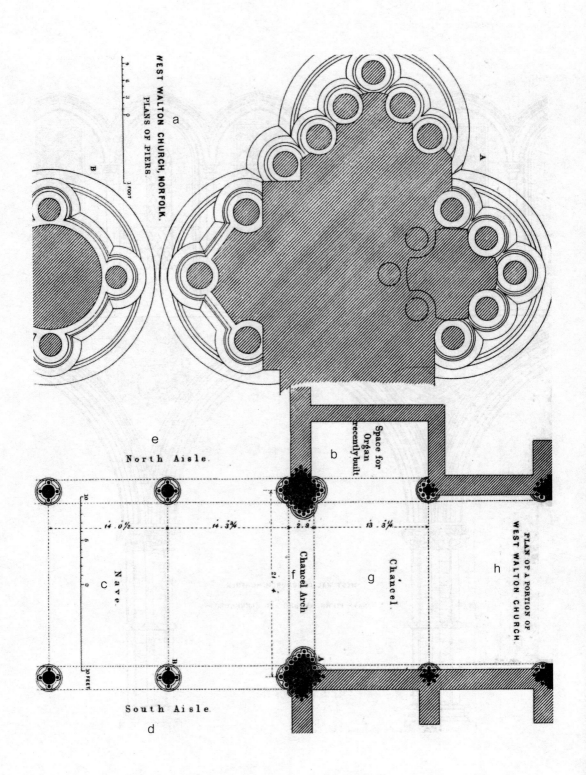

诺福克西沃尔顿教堂：a.集束柱平面　b.新建管风琴空间　c.中殿　d.北走廊　e.南走廊　f.圣坛拱门　g.圣坛　h.西沃尔顿教堂局部平面

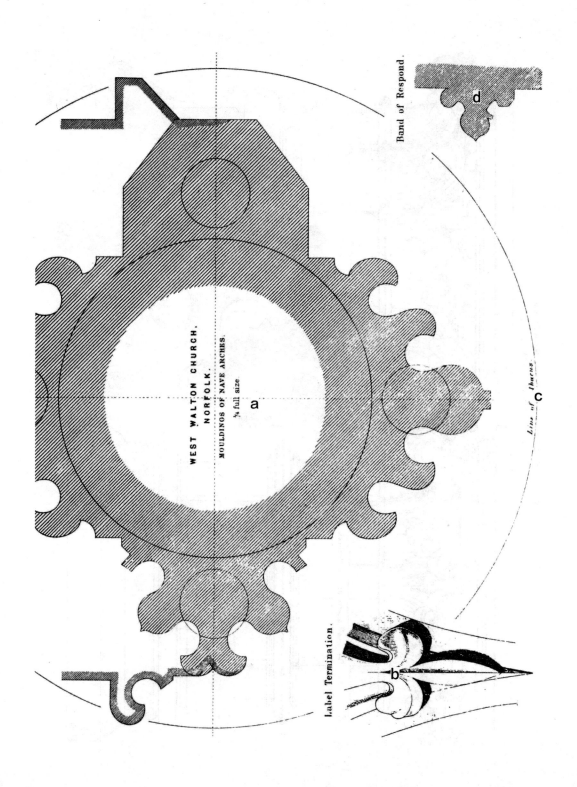

诺福克西沃尔顿教堂：a. 中殿拱线脚　b. 披水石端部　c. 柱顶板线　d. 壁联线脚

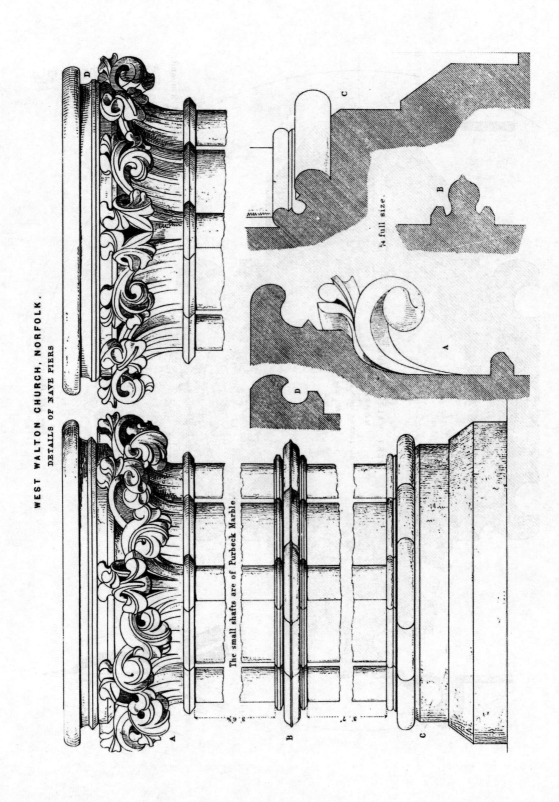

WEST WALTON CHURCH, NORFOLK.
DETAILS OF NAVE PIERS

The small shafts are of Purbeck Marble.

¼ full size.

诺福克西沃尔顿教堂：中殿集束柱细部波贝克大理石柱身

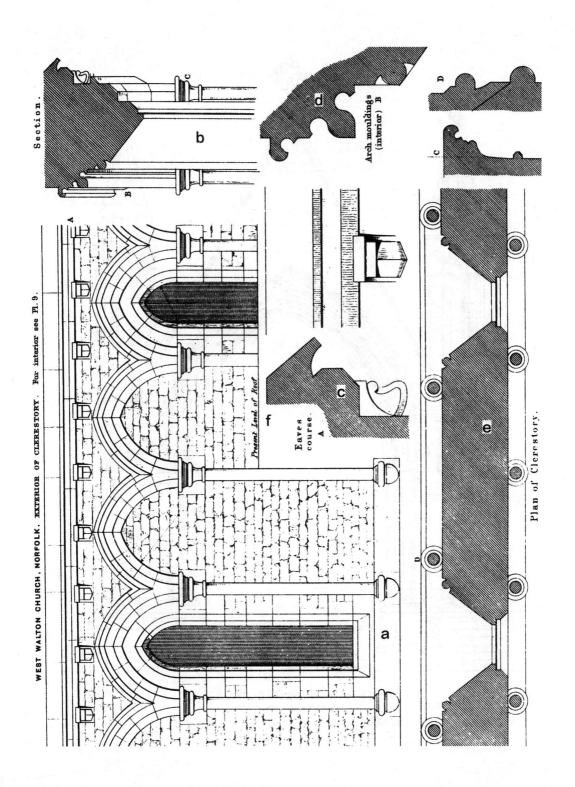

诺福克西沃尔顿教堂：a. 侧高窗外部　b. 剖面　c. 檐口层　d. 内部拱线脚　e. 侧高窗平面　f. 现状屋顶层面

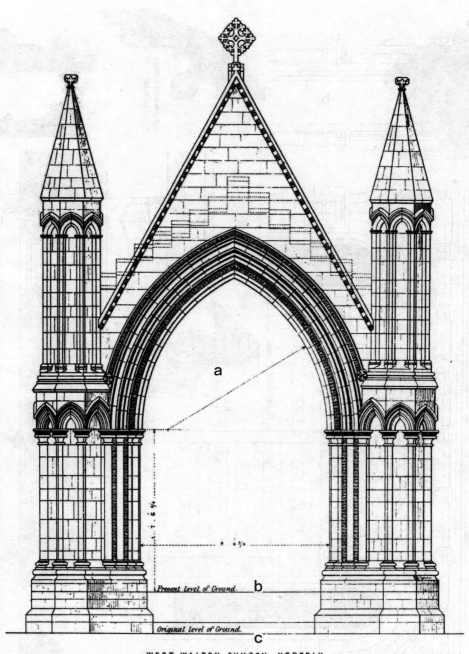

a

b

c

Present level of Ground.

Original level of Ground.

WEST WALTON CHURCH, NORFOLK.

FRONT OF SOUTH PORCH.

*The Gables and the Finials of the Pinnacles are restored.*

诺福克西沃尔顿教堂: a. 南侧门廊正面尖塔的山墙以及被修复的塔尖　b. 现状地平　c. 初始地平

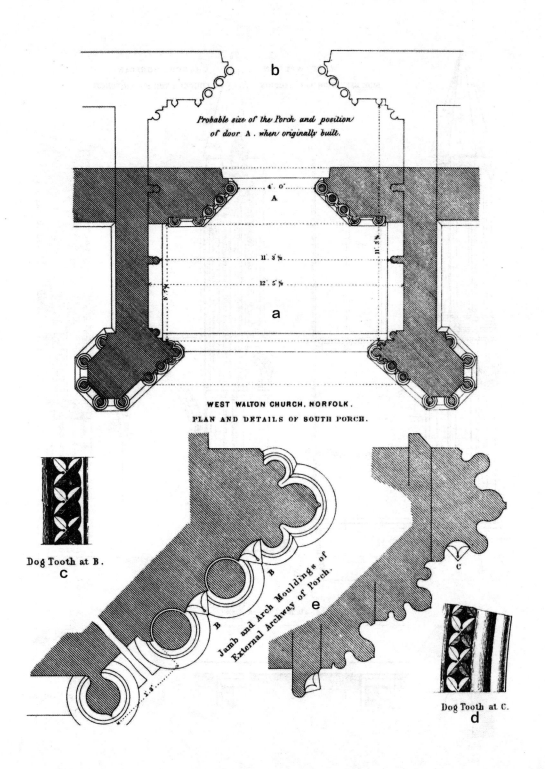

Probable size of the Porch and position
of door A. when originally built.

WEST WALTON CHURCH, NORFOLK.
PLAN AND DETAILS OF SOUTH PORCH.

Dog Tooth at B.
c

Jamb and Arch Mouldings of
External Archway of Porch.

Dog Tooth at C.
d

诺福克西沃尔顿教堂：a. 南侧门廊平面和细部　　b. 原始修建时门廊尺寸及位置推测　　c.B 处四叶花饰　　d.C 处四叶花饰　　e. 门廊外侧拱门侧壁及拱线脚

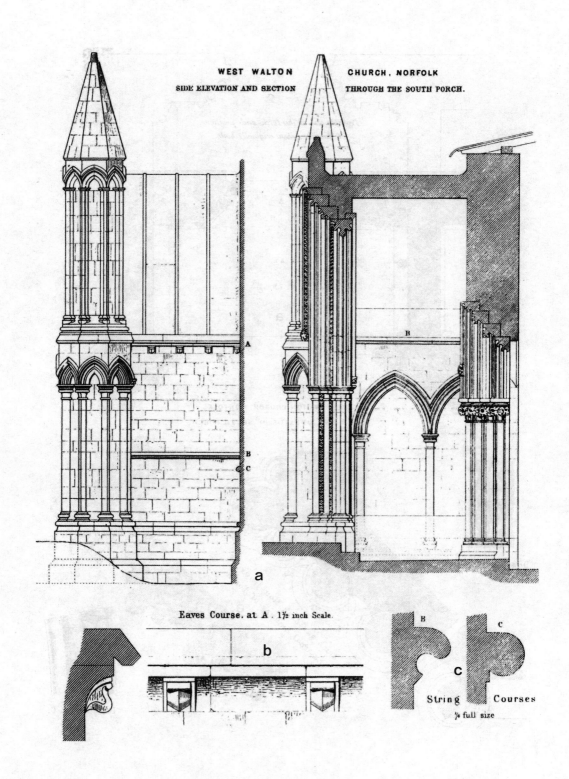

WEST WALTON CHURCH, NORFOLK

SIDE ELEVATION AND SECTION THROUGH THE SOUTH PORCH.

a

Eaves Course. at A . 1½ inch Scale.

b

String Courses
¼ full size

c

诺福克西沃尔顿教堂：a. 侧立面及剖面　b. 檐口层　c. 束带层

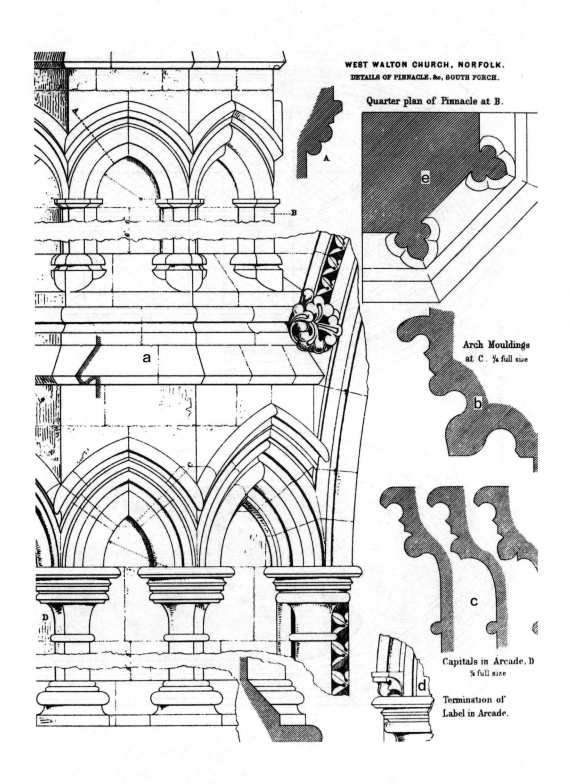

WEST WALTON CHURCH, NORFOLK.
DETAILS OF PINNACLE, &c, SOUTH PORCH.

Quarter plan of Pinnacle at B.

Arch Mouldings
at C . ¼ full size

Capitals in Arcade, D
¼ full size

Termination of
Label in Arcade.

诺福克西沃尔顿教堂：a. 南门廊尖塔细部    b. 拱线脚    c.D 处拱廊内部柱头    d. 拱廊披水石端部    e. B 处四分之一平面

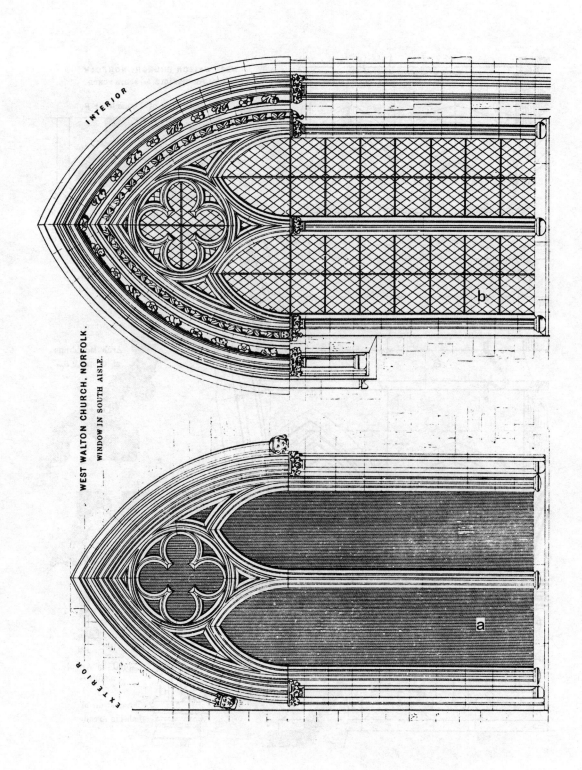

诺福克西沃尔顿教堂南侧走廊窗户：a. 外侧　b. 内侧

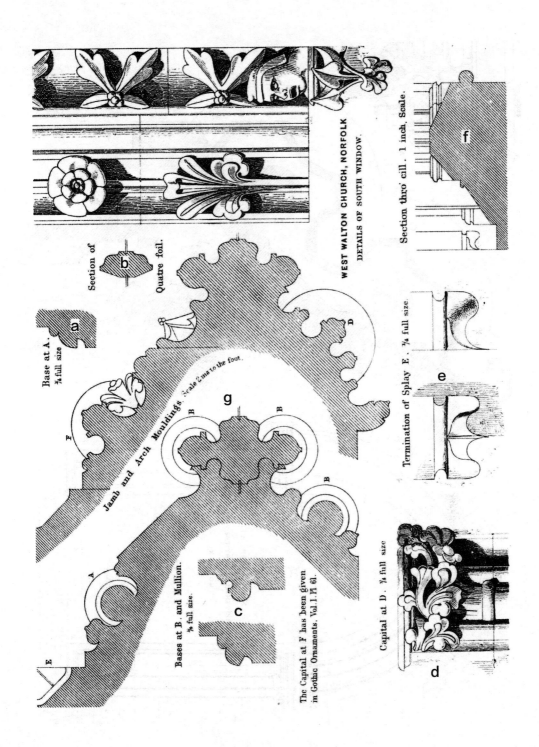

诺福克西沃尔顿教堂南侧窗户细部：a.A处底座　b.四叶饰剖面　c.B处底座及竖框　d.D处柱头　e.斜面E端部　f.门槛剖面　g.侧壁及拱线脚

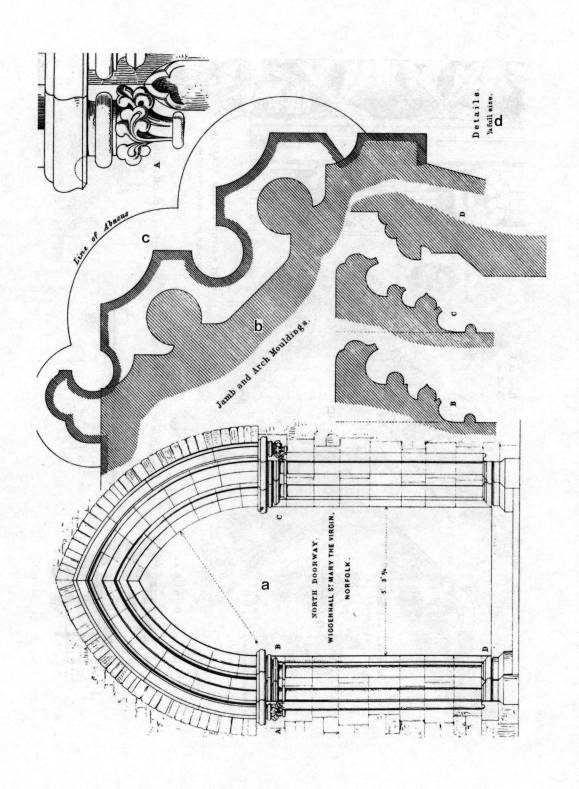

诺福克维金霍尔圣玛丽教堂：a. 北门　b. 侧壁和拱线脚　c. 柱顶板线　d. 细部

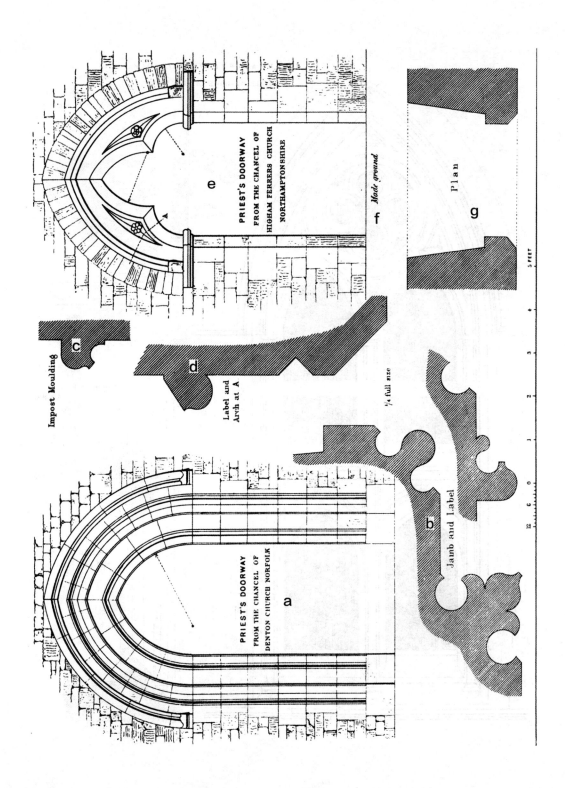

a. 牧师之门　诺福克丹顿教堂圣坛　b. 侧壁和披水石　c. 拱墩线脚　d.A 处披水石和拱

e. 牧师之门　诺坦普顿郡海厄姆费勒斯教堂圣坛　f. 填土　g. 平面

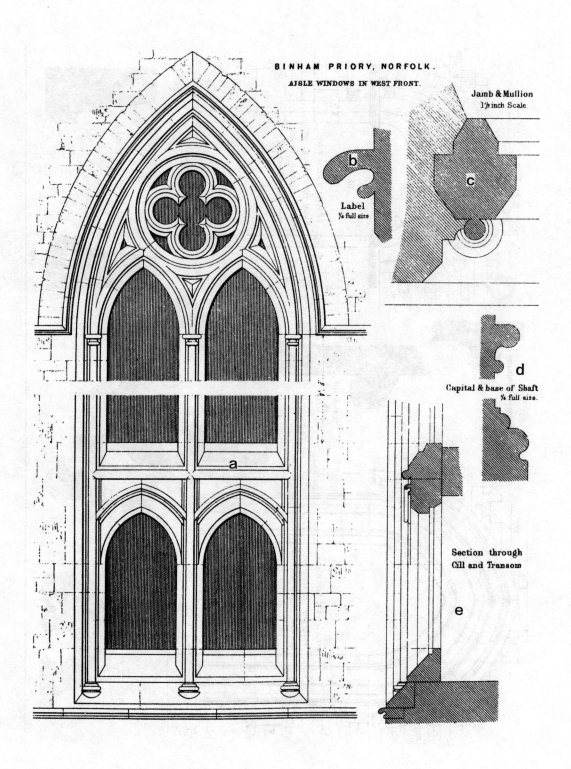

BINHAM PRIORY, NORFOLK.

AISLE WINDOWS IN WEST FRONT.

Jamb & Mullion
1½ inch Scale

Label
¼ full size

b

c

Capital & base of Shaft
¼ full size.

d

Section through
Cill and Transom

e

a

诺福克宾哈姆修道院：a. 西侧廊道窗户　b. 披水石　c. 侧壁和竖框　d. 小束柱柱头和柱础　e. 门槛及横眉剖面

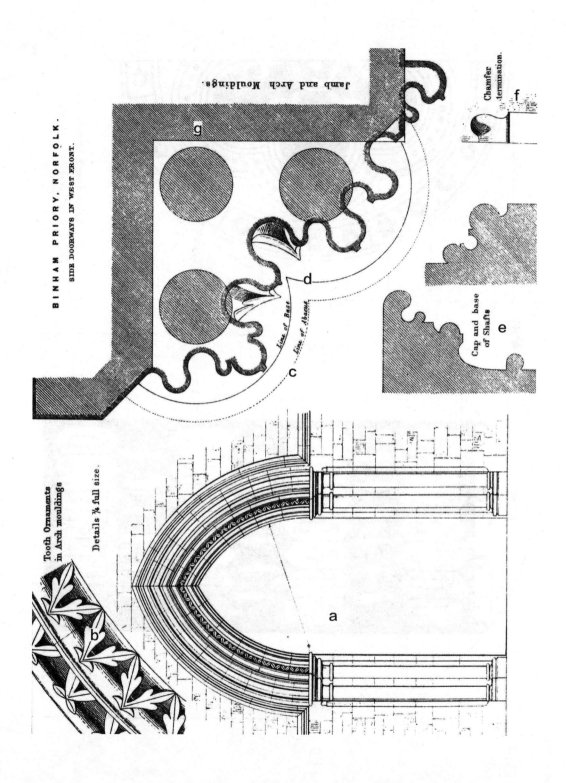

诺福克宾哈姆修道院：a. 西侧侧门　　b. 拱线脚上的四叶花饰细部　　c. 柱顶板线　　d. 柱础外沿线　　e. 柱头和柱础　　f. 凹线端部　　g. 侧壁和拱线脚

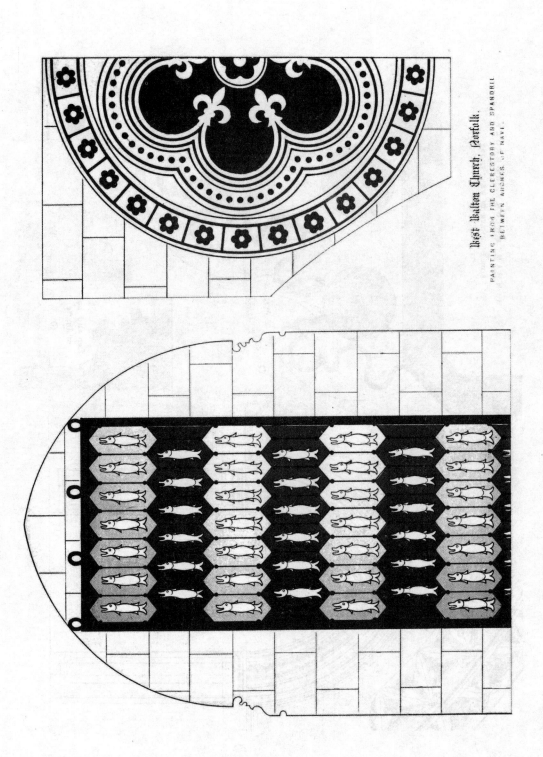

West Walton Church. Norfolk.

PAINTING FROM THE CLERESTORY AND SPANDRIL
BETWEEN ARCHES OF NAVE.

诺福克郡西沃尔顿教堂中殿高窗及拱肩上的彩画

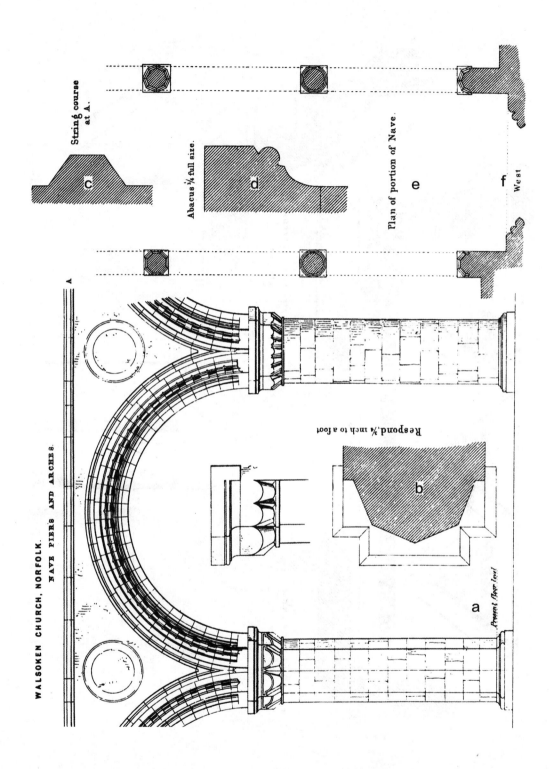

诺福克沃尔索肯教堂中殿集束柱和拱门：a. 现状地平　b. 壁联　c. A 处的束带层剖面　d. 柱顶剖面　e. 中殿局部平面　f. 西侧

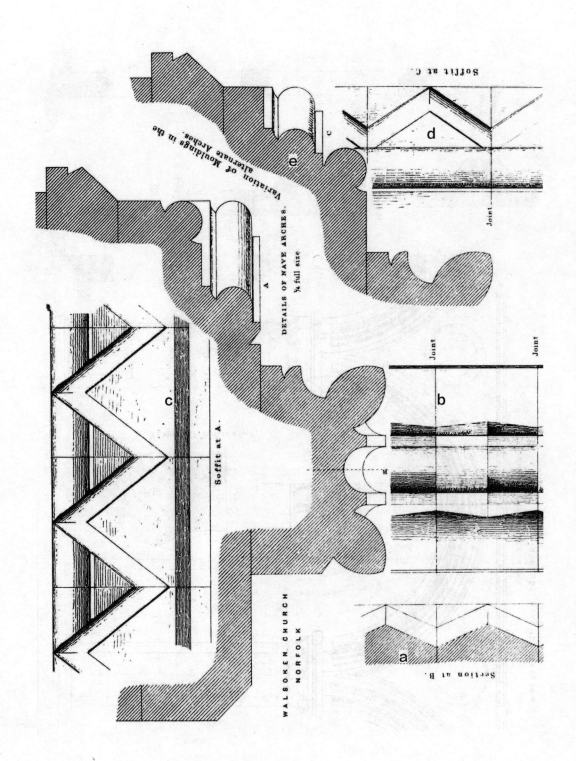

诺福克沃尔索肯教堂中殿拱门细部: a. B 处的剖面　b. 接缝　c. A 处的底面　d. C 处的底面　e. 拱门交替变化的线脚

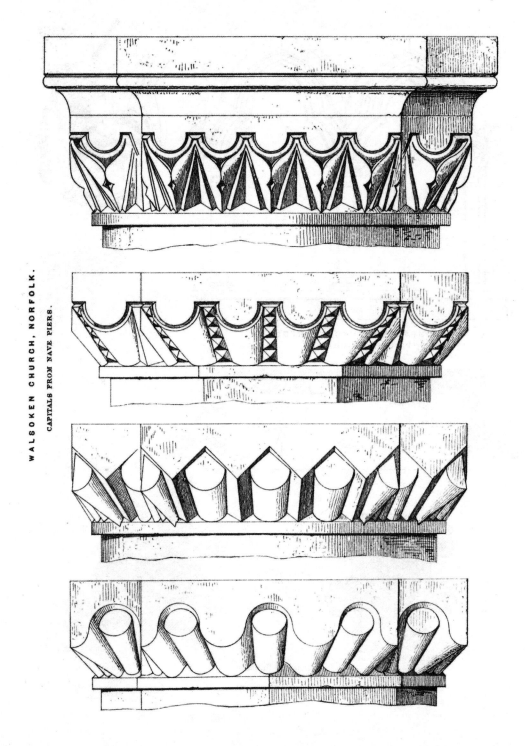

WALSOKEN CHURCH, NORFOLK.

CAPITALS FROM NAVE PIERS.

诺福克沃尔索集束柱柱头

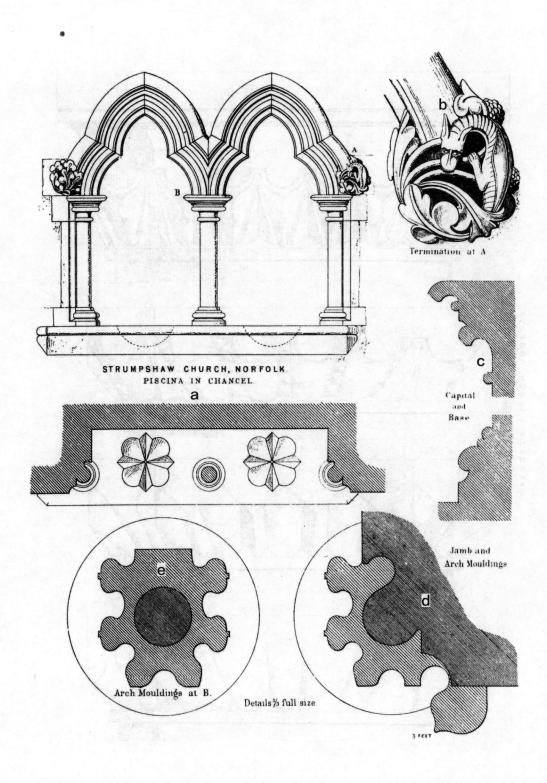

STRUMPSHAW CHURCH, NORFOLK
PISCINA IN CHANCEL.

a

Termination at A

Capital and Base

Jamb and Arch Mouldings

Arch Mouldings at B.

Details ⅓ full size

3 FEET

诺福克斯堂普肖教堂:a. 圣坛里的圣水池　b.A处的端部装饰　c. 柱头和柱础　d. 侧壁和拱脚线　e.B处的拱脚线

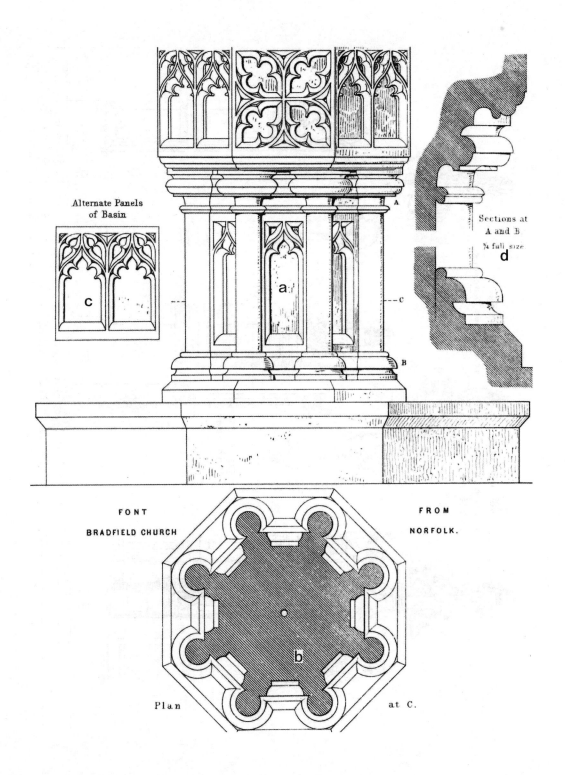

Alternate Panels
of Basin

Sections at
A and B
¼ full size

FONT
BRADFIELD CHURCH

FROM
NORFOLK.

Plan                    at C.

诺福克布拉德菲尔德教堂：a. 洗礼盆    b.C 处的平面    c. 水盆上交替的嵌板    d.A 和 B 处的剖面

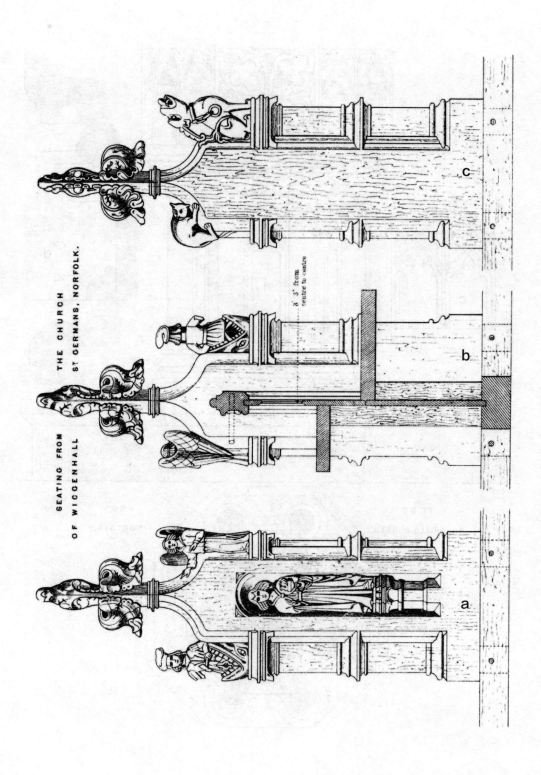

诺福克维金霍尔教堂座椅：a. 中殿长椅端部　b. 长椅剖面　c. 北侧走廊长椅端部

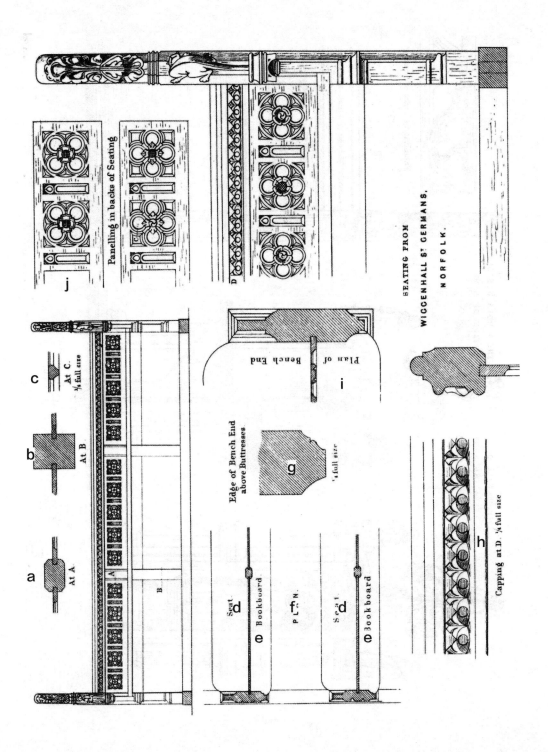

诺福克维金霍尔教堂座椅：a. A处　b. B处　c. C处　d. 座位　e. 木芯板　f. 平面　g. 长椅侧边上部边缘　h. D
处的压顶　i. 长椅端部平面　j. 座椅背部嵌板

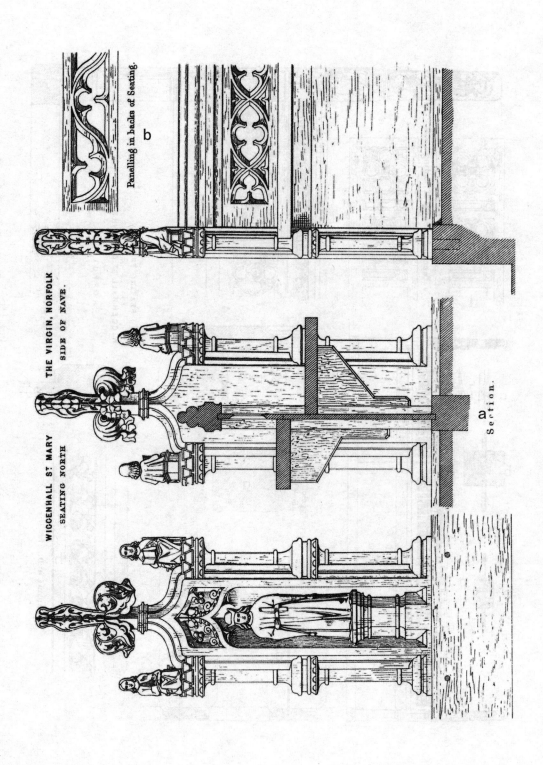

诺福克维金霍尔圣玛丽教堂中殿北侧座椅：a. 剖面    b. 座椅背部嵌板

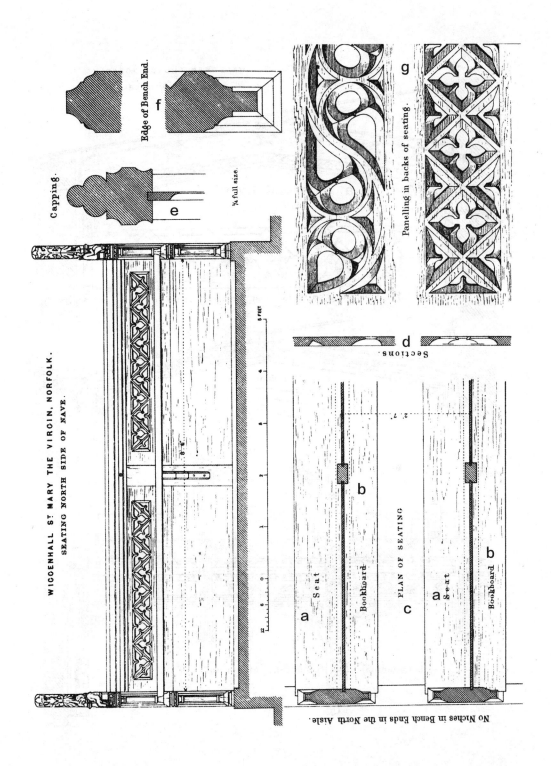

诺福克维金霍尔圣玛丽教堂中殿北侧座椅：a. 座位　b. 木芯板　c. 座椅平面　d. 剖面　e. 压顶　f. 长椅端部边缘　g. 座椅背部嵌板

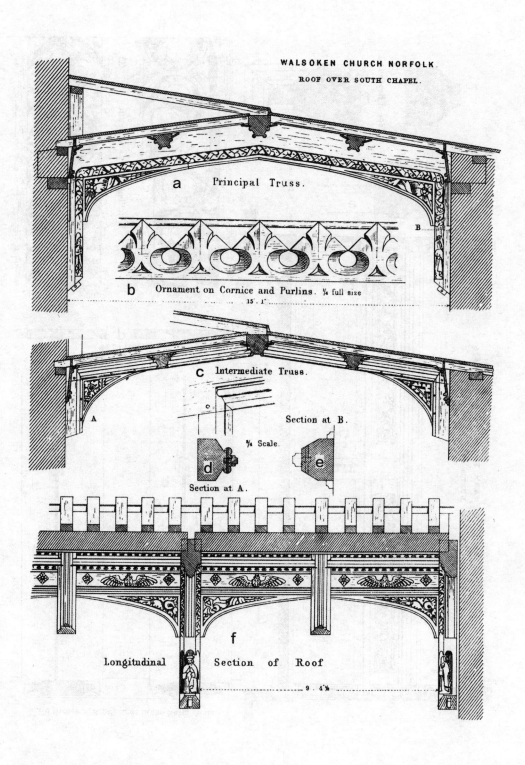

WALSOKEN CHURCH NORFOLK.
ROOF OVER SOUTH CHAPEL.

a   Principal Truss.

b   Ornament on Cornice and Purlins. ¼ full size
15′. 1″

c   Intermediate Truss.

A

Section at B.

¼ Scale.

d   Section at A.   e

B

f

Longitudinal   Section   of   Roof
9. 4 ¾

诺福克沃尔索肯教堂南小礼拜堂屋顶：a. 主屋架　b. 檐口及檩条上装饰　c. 次屋架　d.A 处的剖面　e.B 处的剖面　f. 屋顶纵剖面

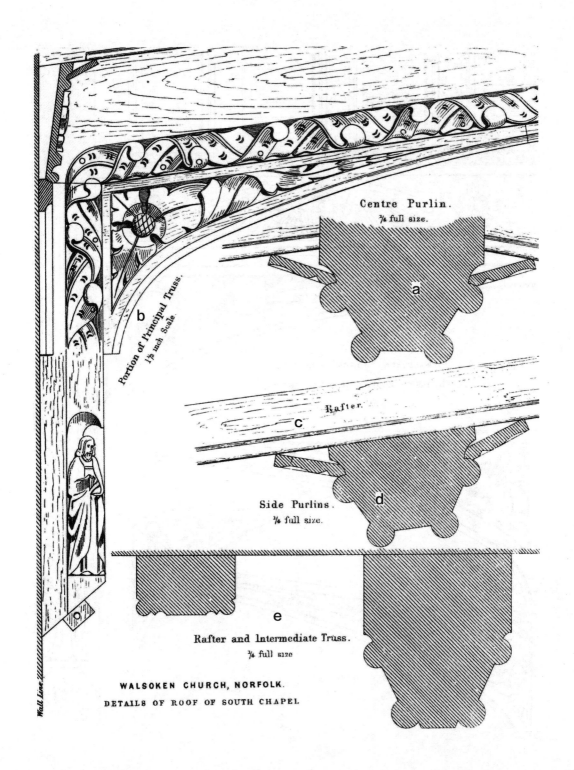

诺福克沃尔索肯教堂南小礼拜堂屋顶细部：a. 中心檩条　b. 主屋架局部　c. 椽子　d. 侧檩条　e. 檩条和次屋架

WALSOKEN CHURCH, NORFOLK.

DETAILS OF ROOF OVER SOUTH CHAPEL.

诺福克沃尔索肯教堂南小礼拜堂屋顶细部

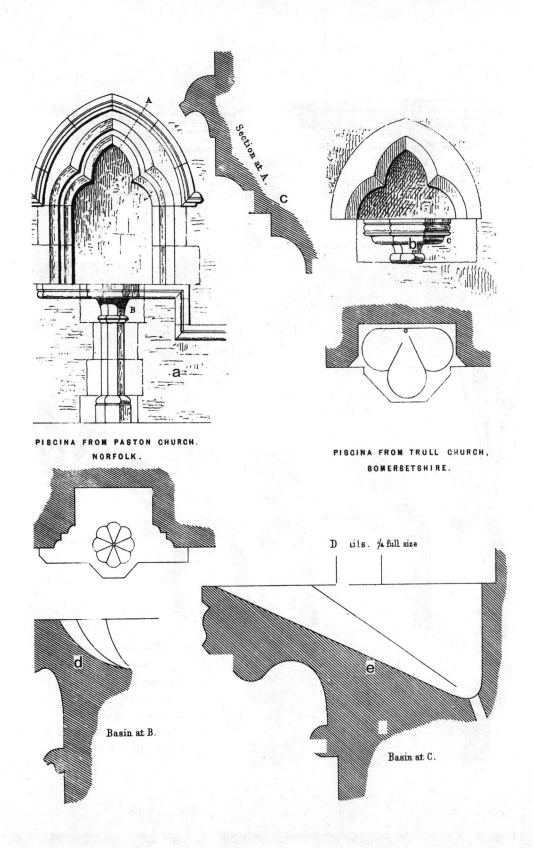

PISCINA FROM PASTON CHURCH,
NORFOLK.

PISCINA FROM TRULL CHURCH,
SOMERSETSHIRE.

Section at A.

Details. ¼ full size

Basin at B.

Basin at C.

a. 诺福克帕斯顿教堂排水石盆　b. 索默塞特郡特鲁尔教堂排水石盆　c.A 处的剖面　d.B 处的盆底　e.C 处的盆底

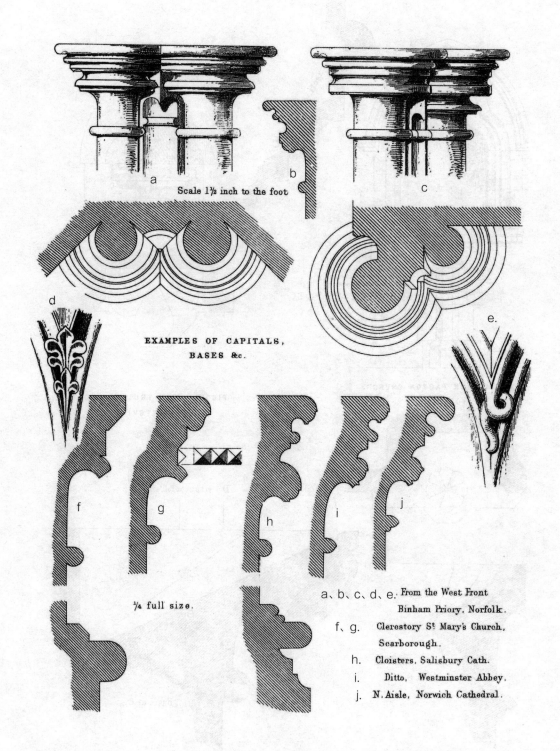

Scale 1½ inch to the foot

EXAMPLES OF CAPITALS,
BASES &c.

¼ full size.

a、b、c、d、e. From the West Front
Binham Priory, Norfolk.

f、g. Clerestory St Mary's Church,
Scarborough.

h. Cloisters, Salisbury Cath.

i. Ditto, Westminster Abbey.

j. N. Aisle, Norwich Cathedral.

柱头和柱础实例：a、b、c、d、e.来自诺福克郡宾哈姆修道院西侧正面　f、g.斯卡伯勒圣玛丽教堂侧高窗　h.索尔兹伯里教堂回廊　i.威斯敏斯特修道院回廊　j.诺威奇教堂北侧廊

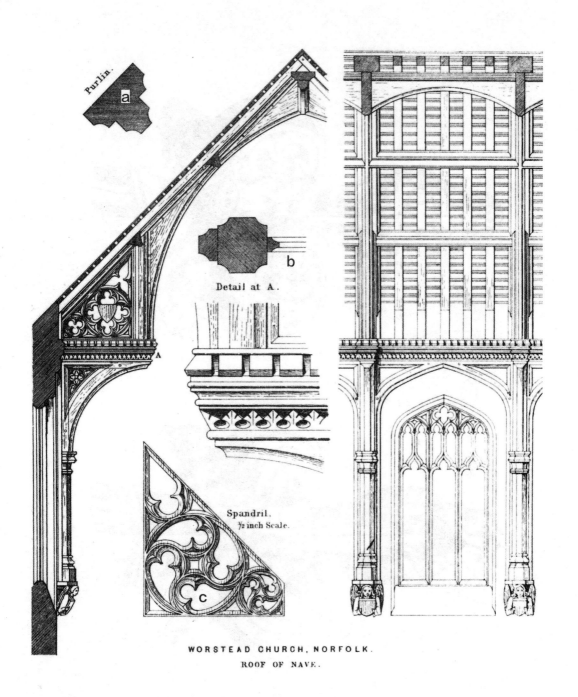

Purlin.

Detail at A.

Spandril.
½ inch Scale.

WORSTEAD CHURCH, NORFOLK.
ROOF OF NAVE.

诺福克沃斯特德教堂中殿屋顶：a.檩条 b.A处的细部 c.拱肩

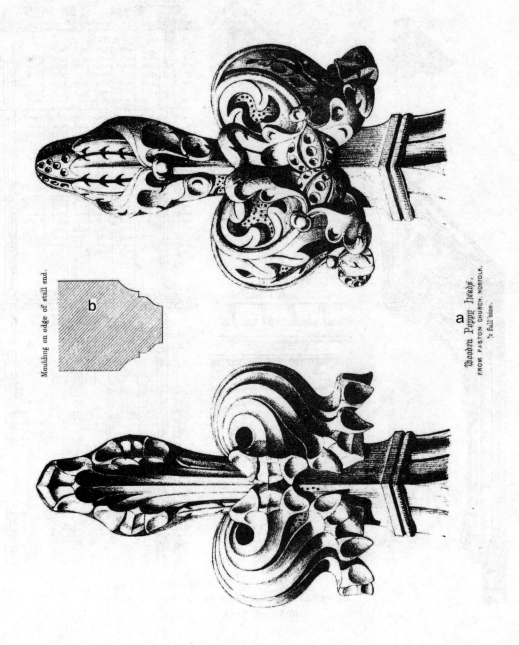

诺福克帕斯顿教堂：a. 木制花冠装饰　b. 座椅端头线脚

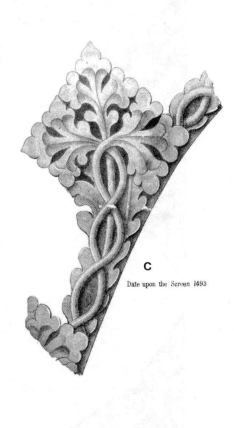

诺福克卢德姆教堂圣坛屏局部展示其通过绘画来变丰富的方式：a. 横楣上铭文局部　　b. 屏风主要竖框局部　c. 大门顶部卷叶形浮雕　　屏风上标注日期为 1493

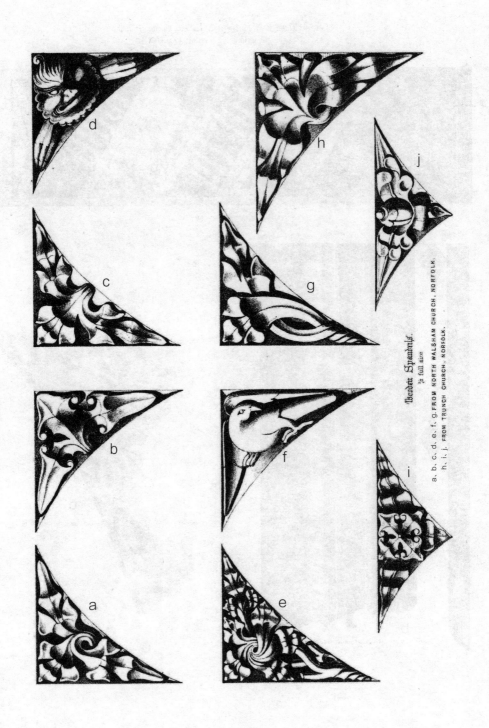

木制拱肩 a、b、c、d、e、f、g. 来自于诺福克北沃尔沙姆教堂    h、i、j. 来源于诺福克创奇教堂

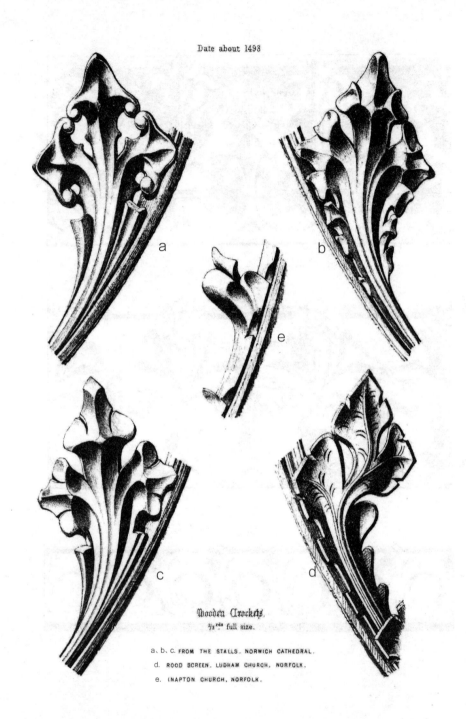

木制卷叶形花饰：a、b、c. 来自于诺维奇教堂座椅　d. 来自于诺福克卢德姆教堂圣坛屏　e. 来自诺福克纳普顿教堂

a

b

c

木制草莓叶冠饰：a. 来自诺福克卢德姆教堂  b. 来自诺福克创奇教堂  c. 来自诺福克纳普顿教堂

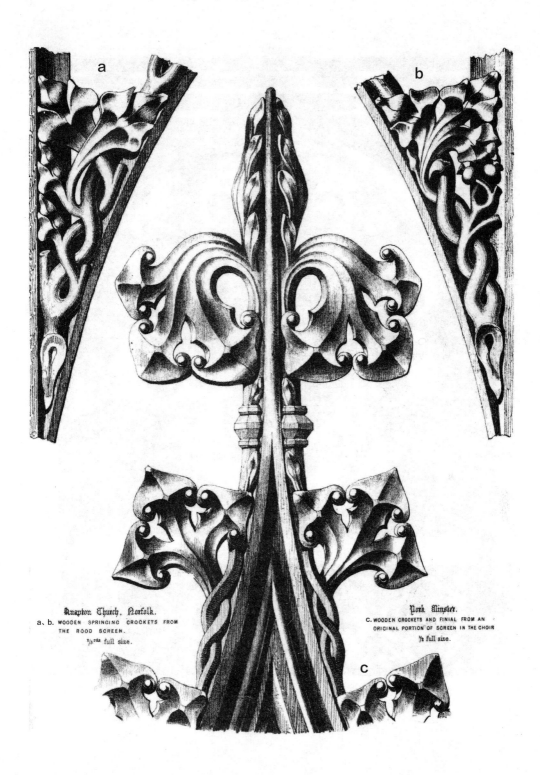

诺福克纳普顿教堂：a、b. 来自于圣坛屏起拱处木制卷叶装饰

约克大教堂：c. 唱诗室内原有屏风木制卷叶装饰及尖顶饰局部

诺福克迪克尔堡教堂圣坛屏上橡木嵌板

诺福克迪克尔堡教堂圣坛屏上橡木拱肩及叶片状尖饰

SPANDRILS BENEATH

HAMMER BEAMS

This roof is illustrated and
further detailed in Mess.rs Brandon's
work on open Timber Roofs

b

Note. The Angel is not in the
position shewn here but
is fixed upon the end of
one of the Hammer beams.
Those at the foot of the Niches
are all lost

c

a

诺福克纳普顿教堂：a. 屋架末尾拱肩和壁龛　　b. 屋顶上有彩画并且在开敞屋顶处画满了梅斯·布兰登的作品
c. 注：天使雕像原本不在所在位置，后来修复时放置在一个悬臂托梁上，原本该在壁龛里的雕像全部丢失

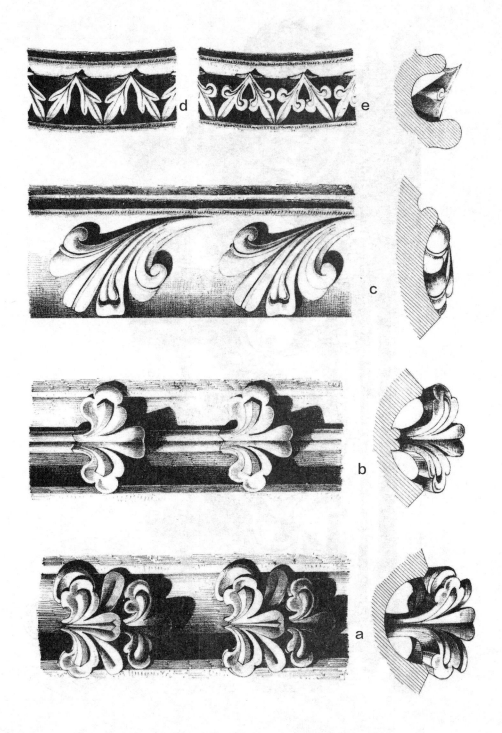

a、b. 以利教堂内柱身间石制装饰　　c. 威斯敏斯特修道院内柱身间石制装饰　　d、e. 诺福克宾哈姆修道院内石制齿形装饰

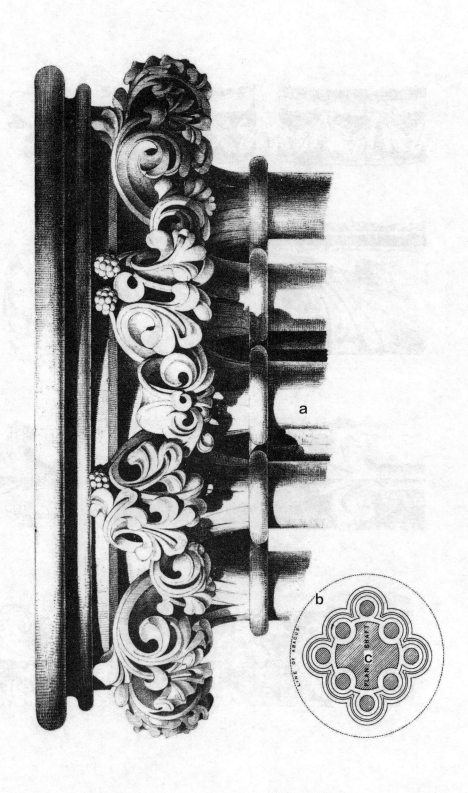

诺福克郡西沃尔顿教堂：a. 圣坛内石制柱头　b. 柱顶板线　c. 柱身平面

诺福克郡西沃尔顿教堂：a. 南侧廊窗户上的石制柱头　b. 柱身平面

诺福克郡西沃尔顿教堂：高窗上花纹装饰彩画——此教堂供奉圣母玛利亚

诺福克郡西沃尔顿教堂高窗上的花纹装饰彩画

诺福克圣尼古拉斯教堂：a. 座椅扶手　　b. 座椅端部边缘线脚

诺福克圣尼古拉斯教堂座椅扶手

诺福克郡圣玛格丽特教堂唱诗室内的木制隔断

诺福克郡圣玛格丽特教堂唱诗室内的木制隔断

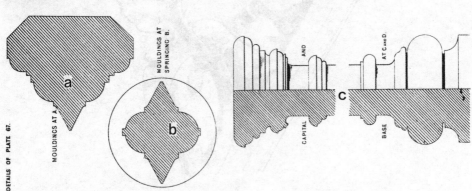

诺福克郡圣玛格丽特教堂唱诗室内木制隔断：a.A 处的线脚　b.B 处的起拱点线脚　c.C 处和 D 处的柱头和柱础　d.E 处的檐口　e.F 处的线脚　f.C 处的小尖塔平面

a

b

诺福克郡东哈林教堂：a.圣坛屏嵌板上带彩绘的橡木装饰　b.A–A处的剖面

诺福克郡圣玛格丽特教堂唱诗室内座椅突出托板

诺福克郡圣玛格丽特教堂唱诗室内座椅突出托板

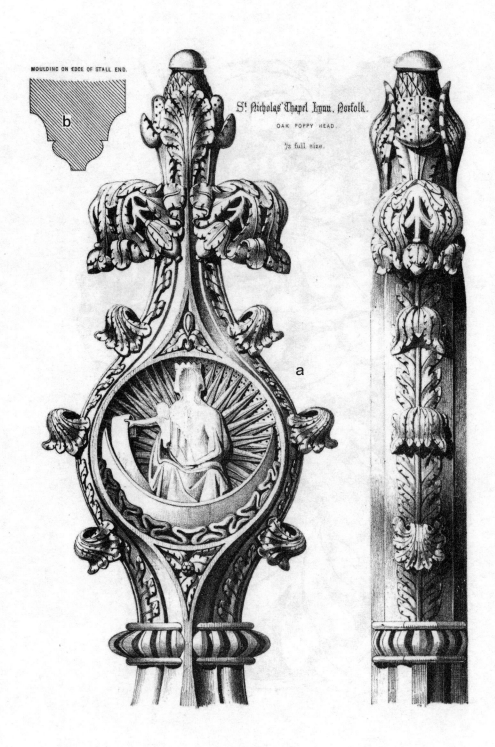

诺福克圣尼古拉斯教堂：a. 橡木罂粟状装饰　b. 座椅端部边缘线脚

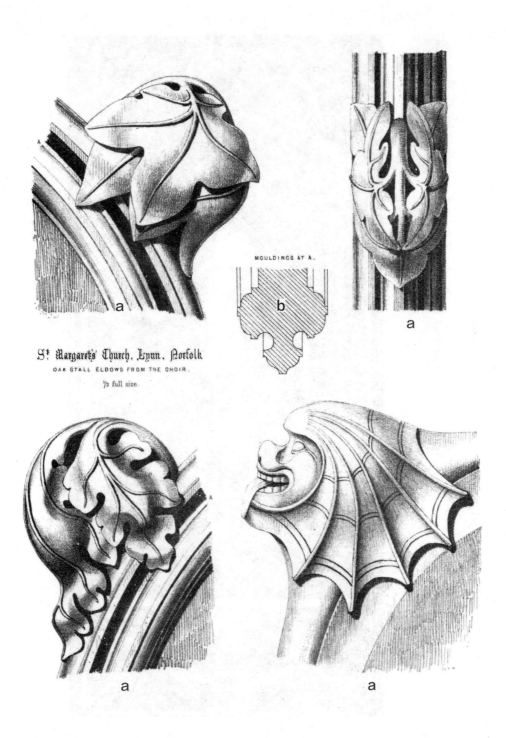

诺福克郡圣玛格丽特教堂：a. 唱诗室内座椅的橡木扶手　　b.A 处的线脚

诺福克郡卡布鲁克教堂圣坛屏上带彩画的橡木嵌板

The Church is dedicated to SS. Paul & Peter.

诺福克郡士瓦弗汉教堂屋顶上象征彼得和保罗及其受难的木制天使像（此教堂供奉彼得和保罗）

Swaffham Church, Norfolk.

WOODEN SPANDRILS FROM THE ROOF

Scale 2 in.° to the foot.

诺福克郡士瓦弗汉教堂屋顶上木制拱肩

ST PETER.

a

Ranworth Church. Norfolk.
PORTIONS OF DIAPER FROM THE DRESSES OF THE
APOSTLES FROM THE ROOD SCREEN
Note._ Yellow indicates Gold.
½ full size.

ST JUDE.

b

ST BARTHOLOMEW.

c

诺福克郡兰沃思教堂圣坛屏上使徒外套上花饰局部：a. 圣彼得　b. 圣朱迪　c. 圣巴塞洛缪

Plain plastered Cove **b**

a

Ranworth Church. Norfolk.

CENTRE PORTION OF ROOD SCREEN.

Scale 1½ inch to the foot.

诺福克郡兰沃思教堂：a. 圣坛屏中心部分　　b. 简单涂以灰泥的凹槽

诺福克郡兰沃思教堂圣坛屏剖面、平面及嵌板：a.阁楼地面　b.拱肋　c.简单涂以灰泥的凹槽　d.暗示出侧面的穹顶　e.横梁下部嵌板　f.圣坛　g.圣餐台原本位置　h.圣坛屏平面

Ranworth Church. Norfolk.

PAINTING FROM THE ROOD SCREEN.

½ full size.

诺福克郡兰沃思教堂圣坛屏上彩画标有 * 号的饰线都是红色的：a. 白色　　b. 横梁下面　　c. 横梁上面　　d. 主竖框　　e.A 处的彩画　　f. 横梁局部　　g. 屏风侧面横梁

来自诺福克的上色装饰：a、d. 来自南伯林厄姆教堂布道台　b、c. 来自兰沃思教堂圣坛屏　e、f. 来自斯特伦普肖教堂圣坛屏　g、h. 来自南伯林厄姆教堂圣坛屏

第4章

# 贝弗利地区
# 教堂建筑

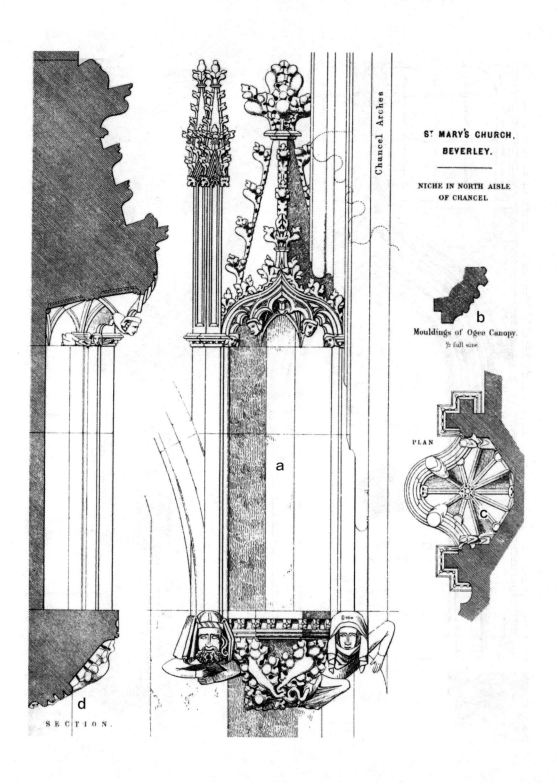

贝弗利圣玛丽教堂：a. 圣坛北部走廊壁龛　b. 二心内心桃尖拱顶棚线脚平面　c. 平面　d. 剖面

JAMB OF DOOR.

Jamb and
Arch Mouldings.
¾ full size.

Side of upper portion of Canopy.

S.T MARY'S CHURCH.
BEVERLEY.

DOORWAY TO CHAPEL
IN CHANCEL.

Capitals · ¼ full size.

PLAN.

贝弗利圣玛丽教堂：a. 通往圣坛小礼拜堂的大门　b. 平面　c. 交叉拱束柱　d. 顶棚侧上侧局部　e. 侧壁和拱
线脚　f. 柱头

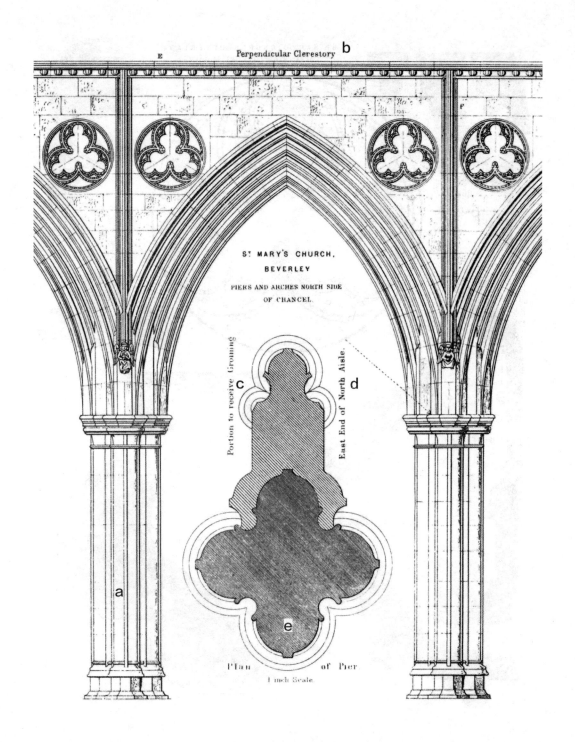

贝弗利圣玛丽教堂：a. 圣坛北部集束柱和拱　b. 垂直高窗　c. 承载交叉拱部分　d. 北侧走廊东部尽端　e. 束柱平面

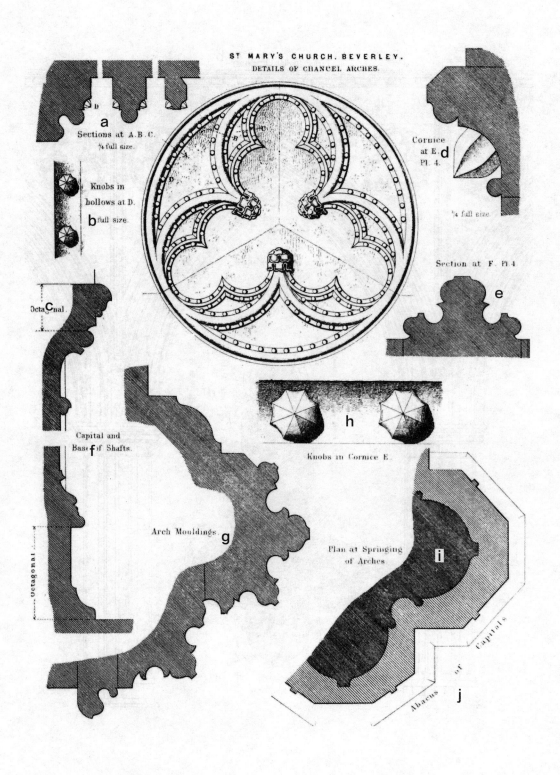

ST MARY'S CHURCH. BEVERLEY.
DETAILS OF CHANCEL ARCHES.

贝弗利圣玛丽教堂圣坛拱细部：a.A、B、C 处的剖面　b.D 处的凹槽处凸起　c.八边形　d.E 处的檐口　e.F
处的剖面　f.束柱柱头及柱础　g.拱线脚　h.E 檐口处的凸起　i.起拱点平面　j.圆柱顶板

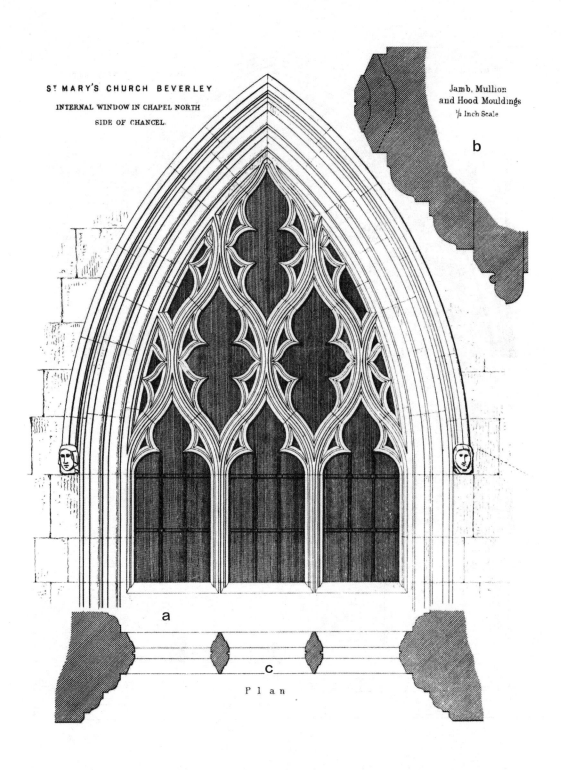

ST MARY'S CHURCH BEVERLEY

INTERNAL WINDOW IN CHAPEL NORTH
SIDE OF CHANCEL.

Jamb, Mullion
and Hood Mouldings
½ Inch Scale

b

a

c

Plan

贝弗利圣玛丽教堂：a. 圣坛北部小礼拜堂内部窗户　b. 侧壁、竖框及出檐线脚　c. 平面

ST MARY'S CHURCH, BEVERLEY.                    EAST WINDOW OF NORTH AISLE OF CHANCEL.

Half Plan

贝弗利圣玛丽教堂 : a. 圣坛北部走廊东侧窗户    b. 中部平面

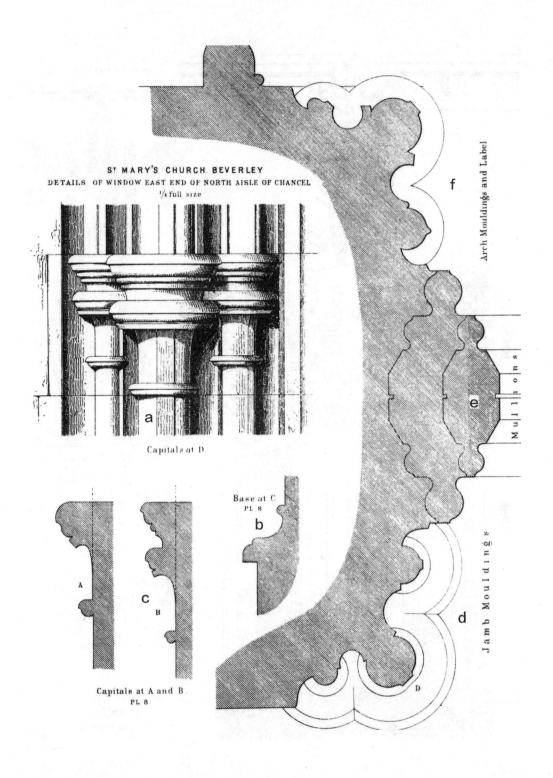

ST MARY'S CHURCH. BEVERLEY
DETAILS OF WINDOW EAST END OF NORTH AISLE OF CHANCEL
¼ full size

Capitals at D

Arch Mouldings and Label

Mullions

Jamb Mouldings

Base at C
PL. 8

Capitals at A and B.
PL. 8

贝弗利圣玛丽教堂圣坛北部走廊东侧尽端窗户细部：a.D 处的柱头　b.C 处的柱础　c.A 和 B 处的柱头　d. 侧壁线脚　e. 竖框　f. 拱线脚和披水石

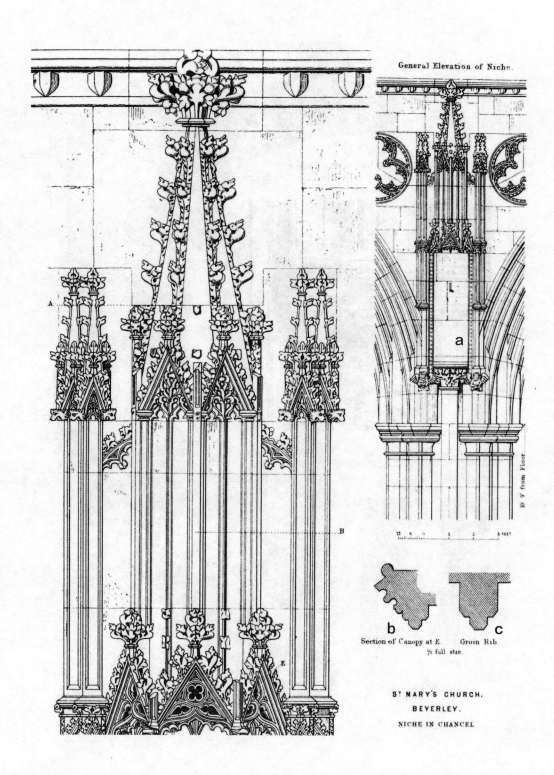

General Elevation of Niche.

a

10.9 from Floor.

12 6 0 1 2 3 FEET

b
Section of Canopy at E.
½ full size.

c
Groin Rib.

St MARY'S CHURCH,
BEVERLEY.
NICHE IN CHANCEL

贝弗利圣玛丽教堂圣坛内壁龛：a. 壁龛总体立面　b.E 处的顶部剖面　c. 穹棱肋

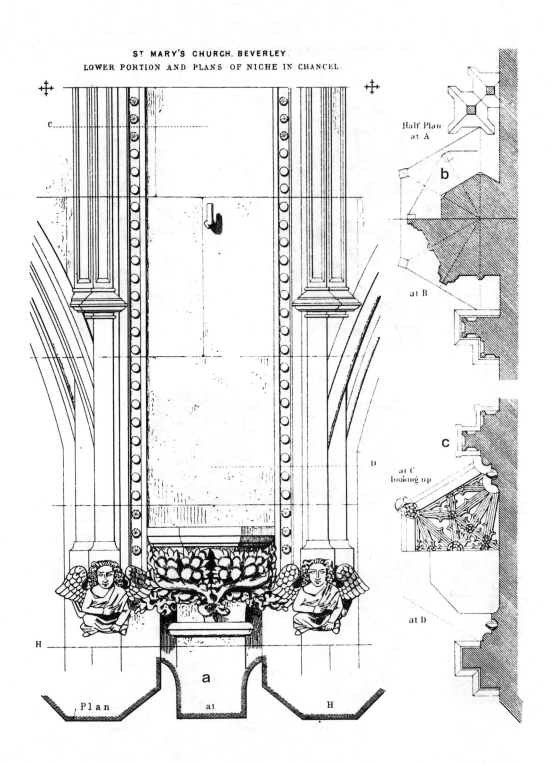

贝弗利圣玛丽教堂圣坛内部壁龛低处局部及平面：a.H 处的平面　　b.A 处的半平面　　c.C 处的仰视

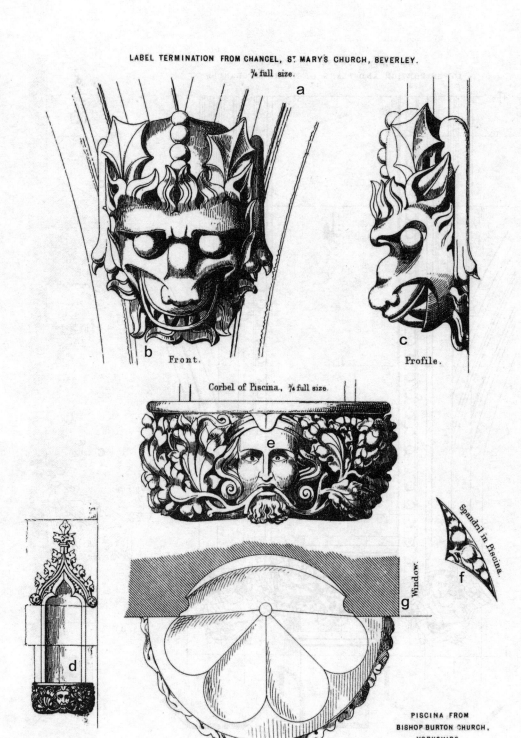

LABEL TERMINATION FROM CHANCEL, ST. MARY'S CHURCH, BEVERLEY.
¾ full size.

a

b  Front.

c  Profile.

Corbel of Piscina, ¾ full size.

e

d

Window.

g

Spandril in Piscina.

f

PISCINA FROM
BISHOP BURTON CHURCH,
YORKSHIRE.

贝弗利圣玛丽教堂：a. 披水石端部  b. 正面  c. 侧面

约克郡伯顿主教堂：d. 排水石盆  e. 排水石盆梁托  f. 排水石盆拱肩  g. 窗户

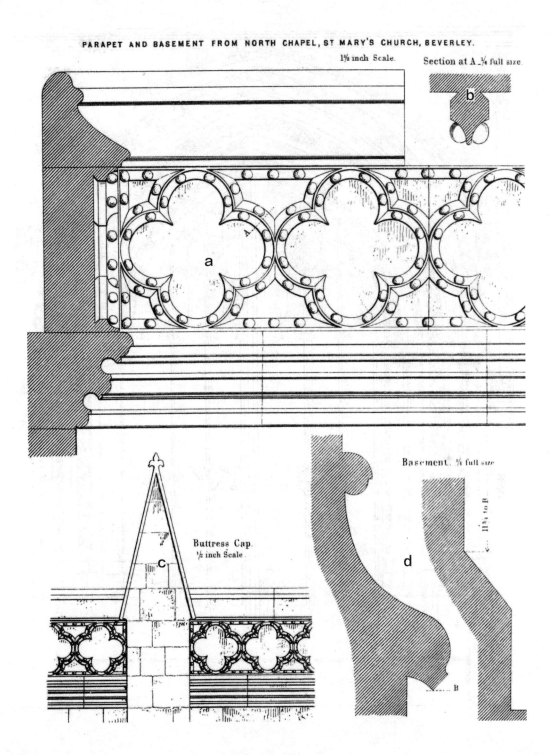

PARAPET AND BASEMENT FROM NORTH CHAPEL, ST MARY'S CHURCH, BEVERLEY.

1½ inch Scale.

Section at A ¼ full size.

Buttress Cap.
½ inch Scale.

Basement. ¼ full size

贝弗利圣玛丽教堂：a. 北侧礼拜堂女儿墙和墙脚　b.A 处的剖面　c. 墩帽　d. 墙脚

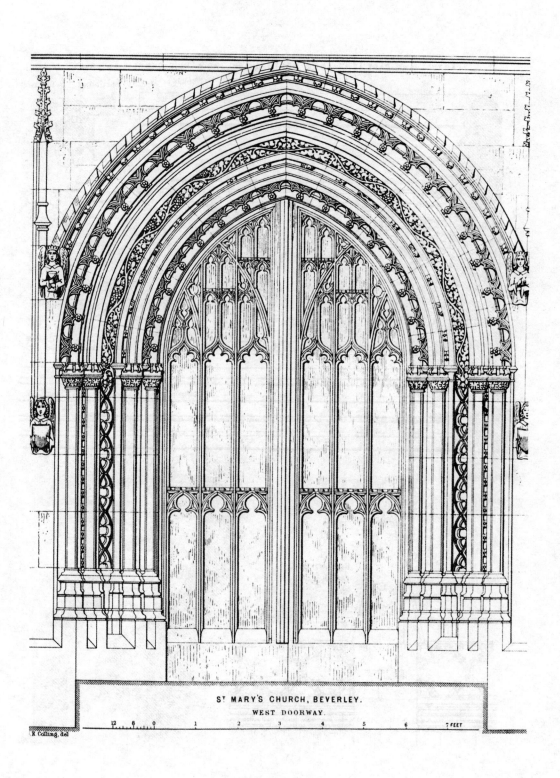

ST MARY'S CHURCH, BEVERLEY.

WEST DOORWAY.

贝弗利圣玛丽教堂西侧大门

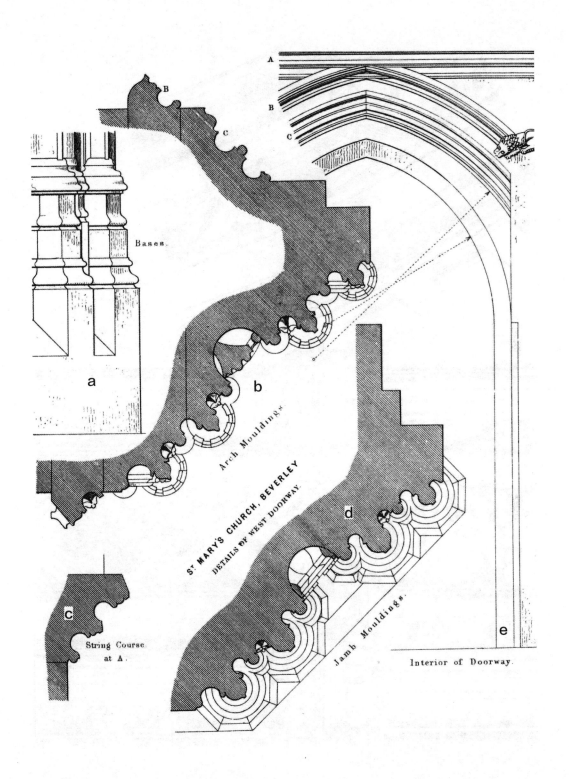

贝弗利圣玛丽教堂：a. 柱础　　b. 拱线脚　　c.A 处的束带层　　d. 柱子线脚　　e. 大门内部

贝弗利圣玛丽教堂西侧大门细部

Capital. 2 inch Scale.

Base. 2 inch Scale.

Label.
¼ full size.

Plan of Pier and Arch Mouldings
1 inch Scale.

Section at A. 1 inch Scale.

ST MARY'S CHURCH, BEVERLEY.
NAVE PIERS AND ARCHES – NORTH SIDE.

贝弗利圣玛丽教堂：a. 中殿集束柱和拱（北侧）　b. 集束柱平面和拱线脚　c. 柱头　d. 披水石　e. 柱础　f. A处的剖面

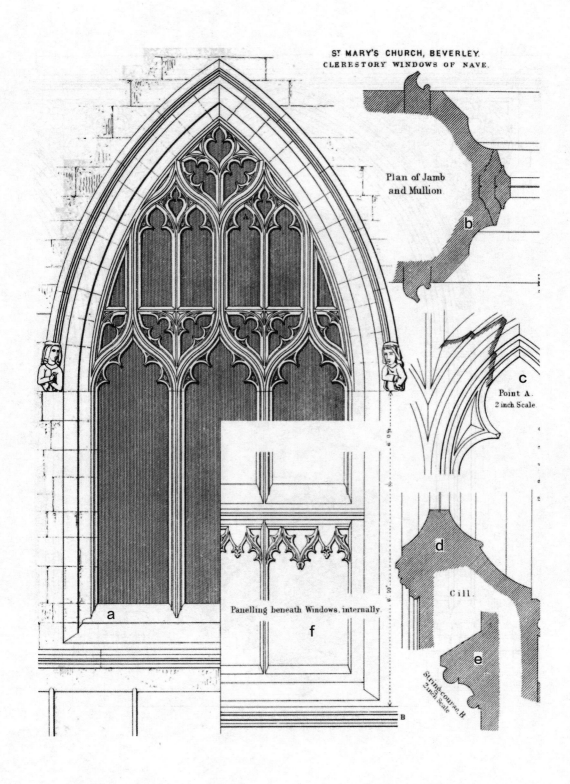

ST. MARY'S CHURCH, BEVERLEY.
CLERESTORY WINDOWS OF NAVE.

Plan of Jamb and Mullion

Point A.
2 inch Scale

Cill

String course B.
2 inch Scale

Panelling beneath Windows, internally.

贝弗利圣玛丽教堂：a. 中殿高窗　b. 侧壁和竖框　c. A 处　d. 窗台　e. 束带层　f. 窗户内部下方镶板

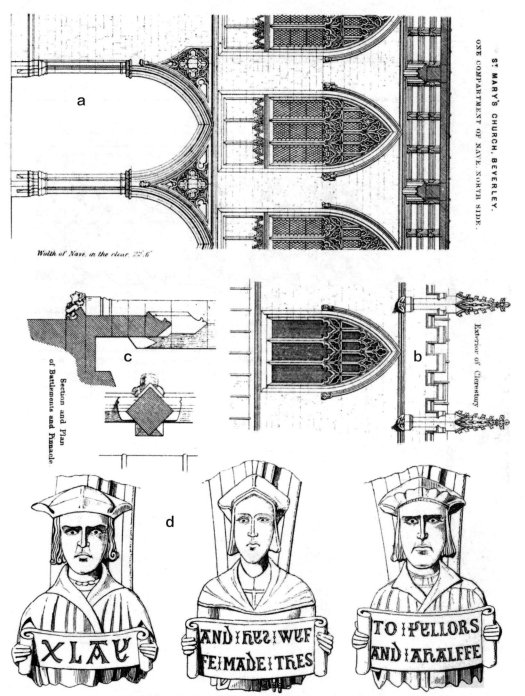

贝弗利圣玛丽教堂：a. 中殿北侧廊一跨　b. 高窗外部　c. 雉堞墙、小尖塔剖面和平面　d. 中殿北部披水石端部

Block A.
2 inch Scale.

Block from St Mary's Church. Scarborough.
2 inch Scale.

a. 贝弗利大教堂女儿墙和檐口墙　b. A 处的砌块　c. 斯卡伯勒圣玛丽教堂砌块

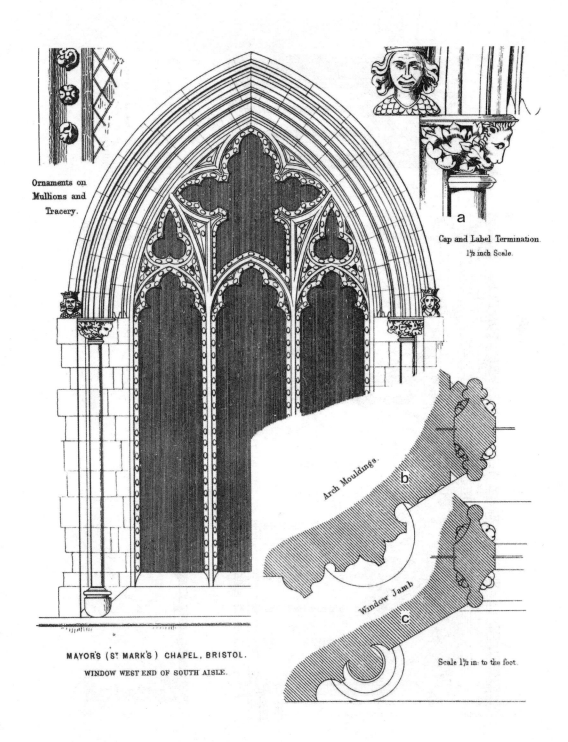

布里斯托尔梅厄教堂南侧廊西尽端窗户：a.柱头和披水石端部    b.拱线脚    c.窗户侧壁

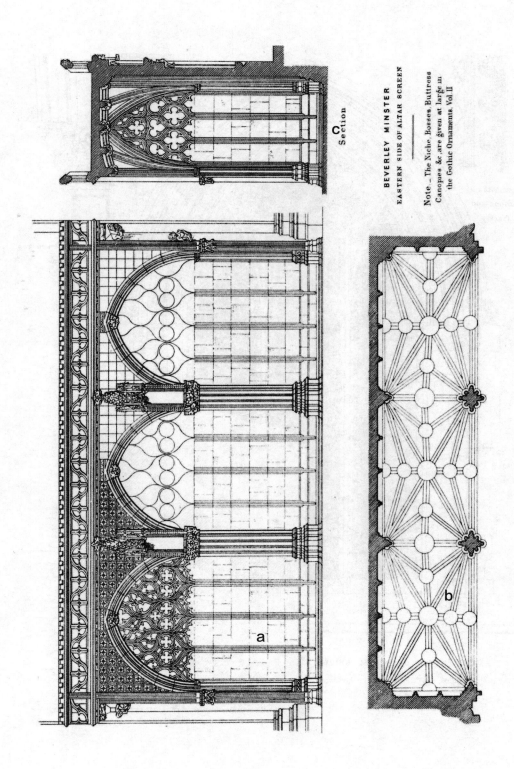

贝弗利大教堂：a. 圣坛隔扇东侧    b. 平面    c. 剖面

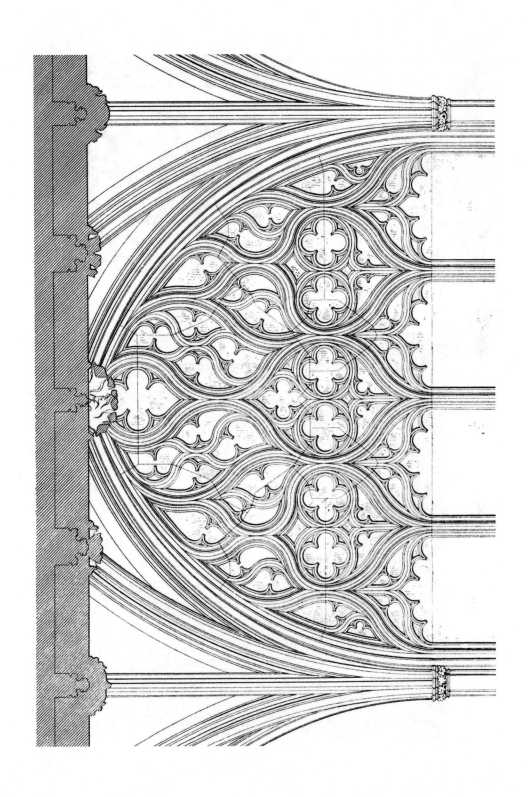

贝弗利大教堂圣坛隔扇东侧窗饰及穹棱

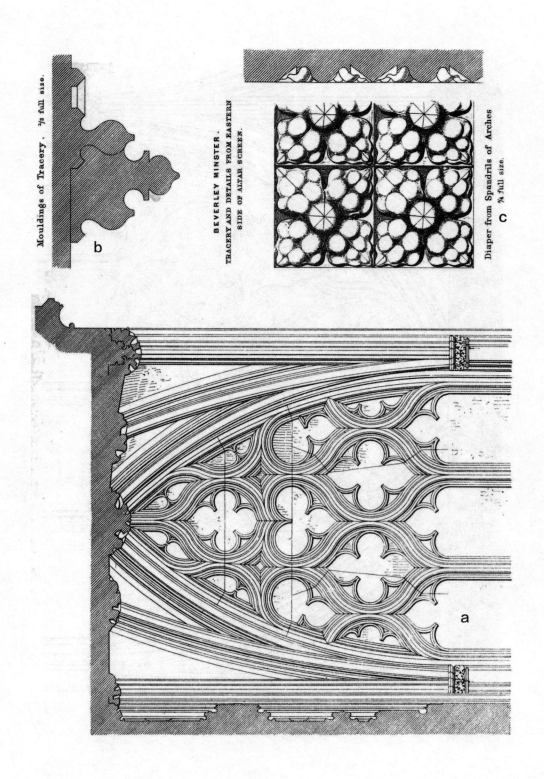

Mouldings of Tracery. ⅔ full size.

b

BEVERLEY MINSTER.
TRACERY AND DETAILS FROM EASTERN
SIDE OF ALTAR SCREEN.

Diaper from Spandrils of Arches ¾ full size.

c

a

贝弗利大教堂：a.圣坛隔扇东侧窗饰及细部　b.窗饰线脚　c.拱肩菱形装饰

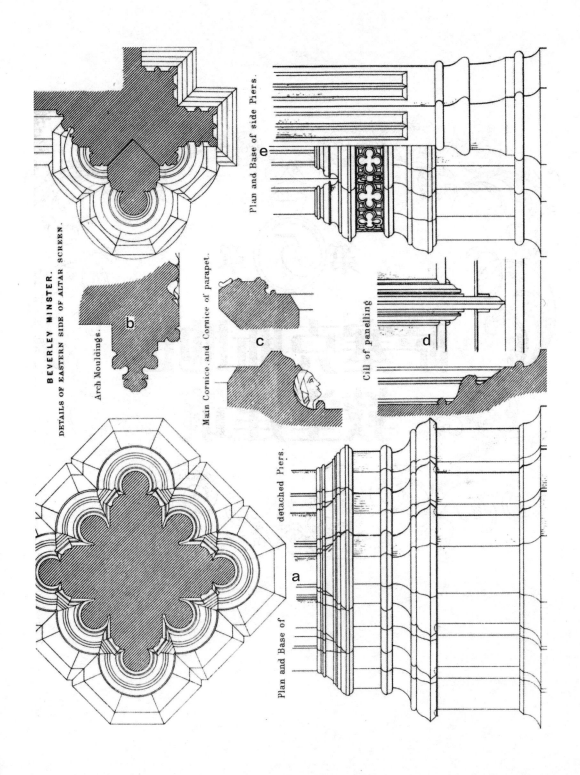

贝弗利大教堂圣坛隔扇东侧细部：a. 独立支墩平面及柱础　b. 拱线脚　c. 主檐口及女儿墙檐口　d. 嵌板底部　e. 侧壁平面及柱础

第5章

# 萨塞克斯地区教堂建筑

From Piscina in Chancel

萨塞克斯郡温切尔西教堂中位于圣殿排水石盆上的石制棱形装饰

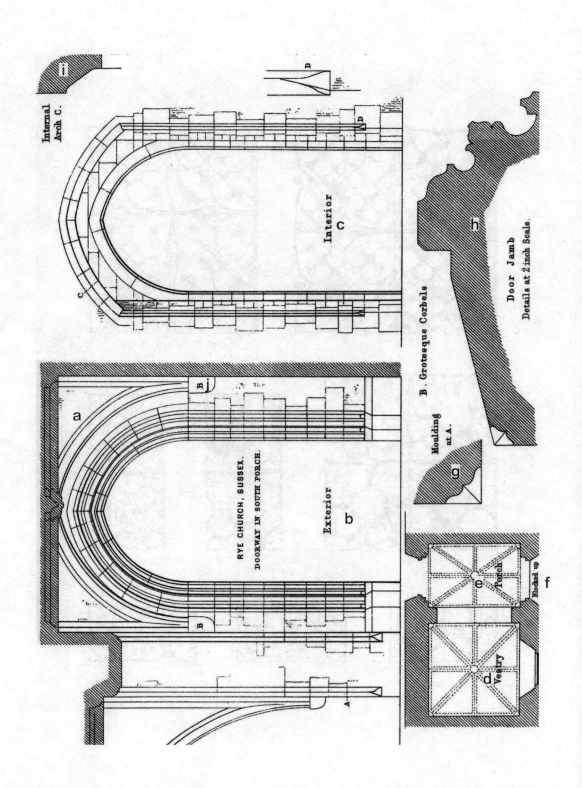

萨塞克斯郡赖伊教堂：a. 南侧门廊大门　b. 外侧　c. 内侧　d. 法衣室　e. 门廊　f. 被封堵　g.A 处线　h. 门侧墙　i. 内侧拱 C　j. 异形托梁

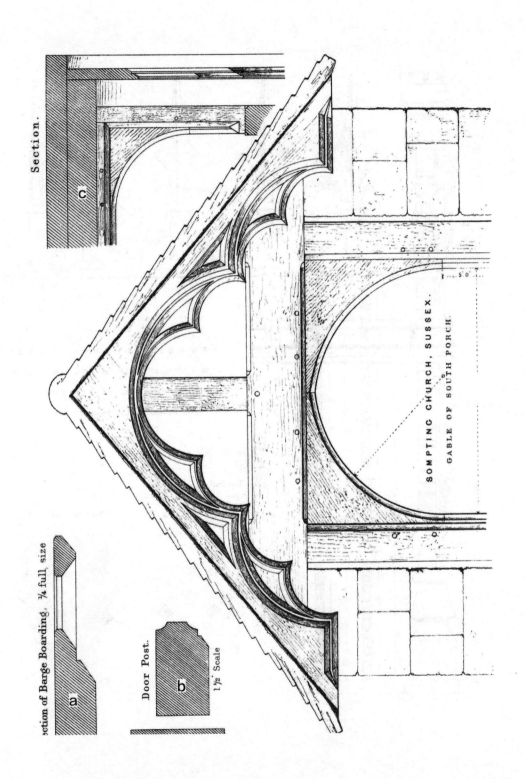

萨塞克斯郡桑普廷教堂南门廊山墙：a. 挡风板剖面　b. 门柱　c. 剖面

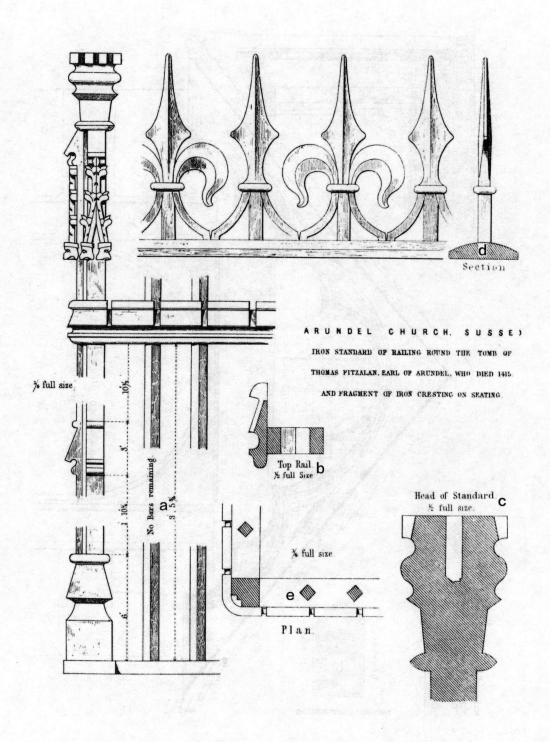

萨塞克斯郡阿伦德尔教堂环绕托马斯·菲查伦（阿伦德尔伯爵，死于 1415 年）墓地的栏杆的铁支柱以及座椅
上铁制顶饰碎片：a.无存留铁条　b.顶部铁条平面　c.铁支柱顶部　d.剖面　e.平面

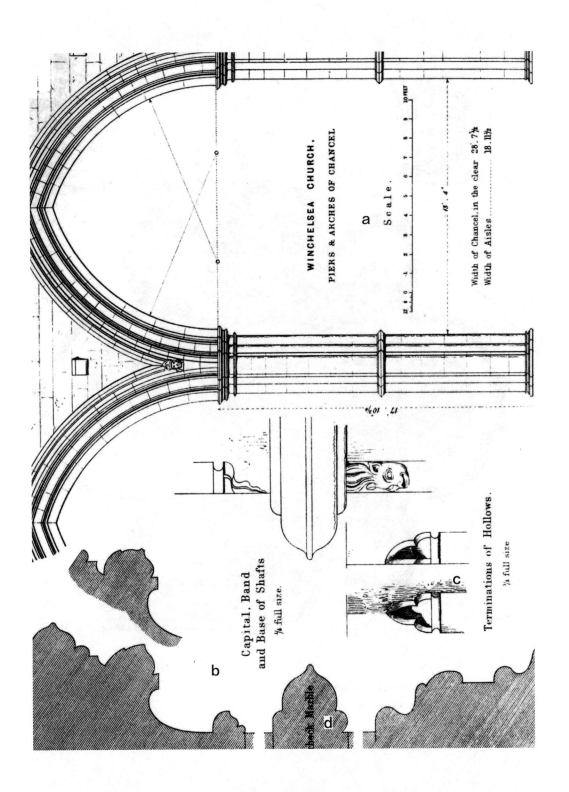

WINCHELSEA CHURCH.

PIERS & ARCHES OF CHANCEL.

a

Scale.

Width of Chancel, in the clear 28'.7½
Width of Aisles ............ 18'.11½

Capital, Band
and Base of Shafts
¼ full size.

b

Terminations of Hollows.
¼ full size

c

Purbeck Marble

d

温切尔西教堂：a. 圣坛集束柱和拱　b. 束柱柱头、束带和柱础　c. 凹槽端部　d. 波贝克大理石

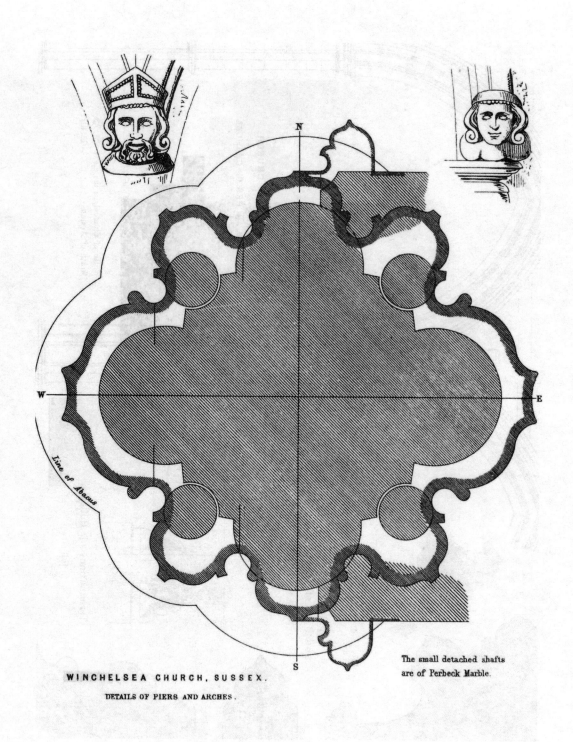

WINCHELSEA CHURCH, SUSSEX.

DETAILS OF PIERS AND ARCHES.

The small detached shafts are of Perbeck Marble.

萨塞克斯郡温切尔西教堂集束柱拱细部、小的独立柱材料为波贝克大理石

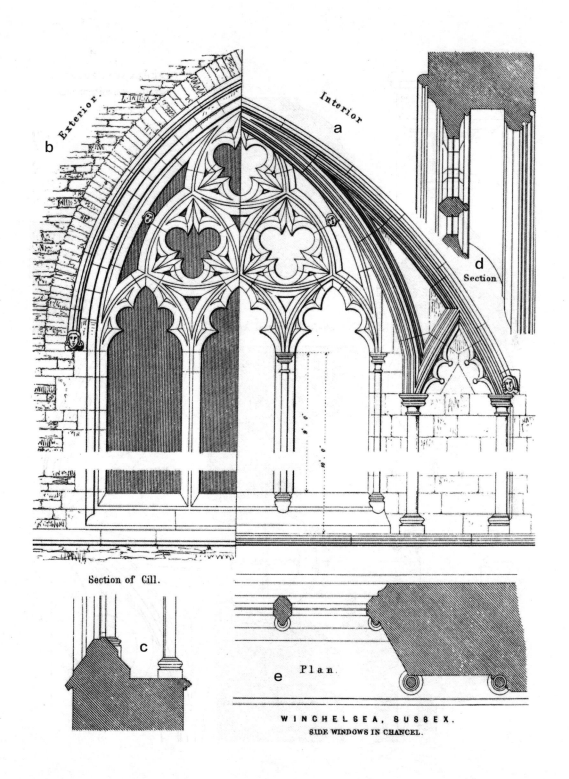

萨塞克斯郡温切尔西教堂圣坛侧窗：a. 内侧　b. 外侧　c. 窗台剖面　d. 剖面　e. 平面

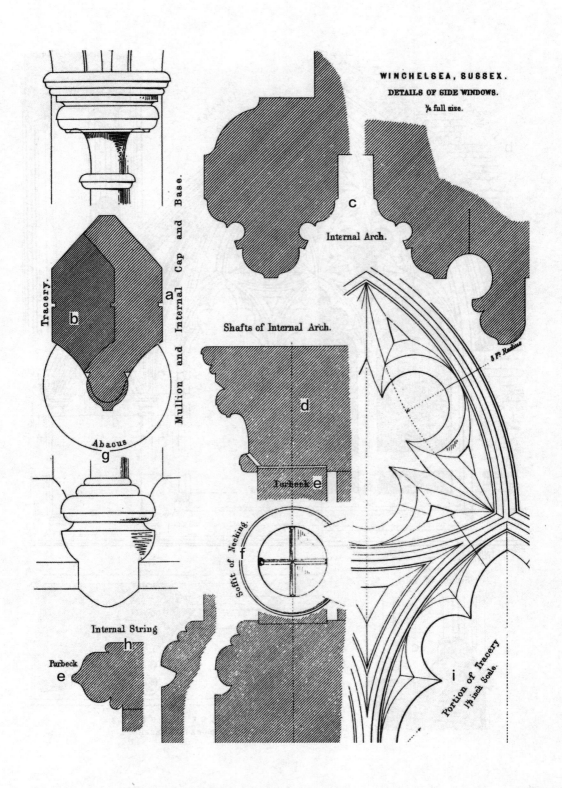

萨塞克斯郡温切尔西教堂侧窗细部：a. 竖框及内部柱头和柱础　b. 窗饰　c. 内部拱　d. 内部拱柱子　e. 波贝克大理石　f. 柱颈线脚　g. 柱顶板　h. 内侧束带　i. 窗饰局部

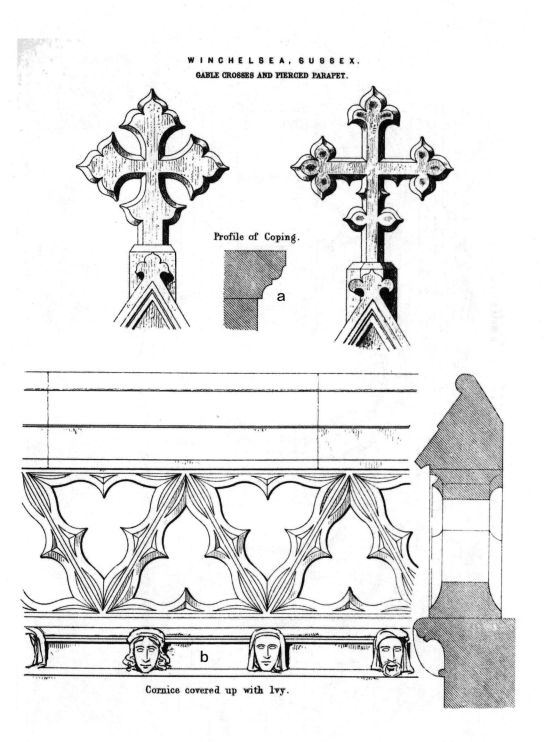

萨塞克斯郡温切尔西教堂山墙十字架和镂空女儿墙：a. 压顶板轮廓　b. 常春藤装饰覆盖的檐口

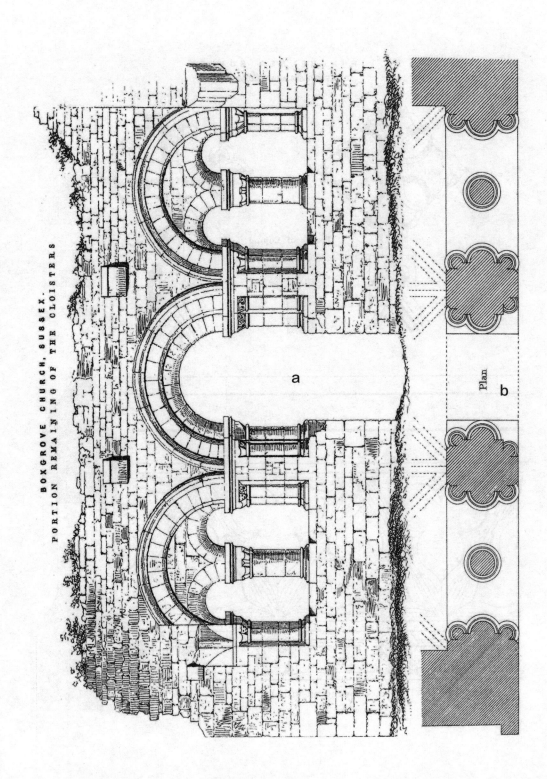

BOXGROVE CHURCH, SUSSEX.
PORTION REMAINING OF THE CLOISTERS.

Plan

a

b

萨塞克斯郡伯克斯格罗夫教堂：a. 遗留的回廊局部　b. 平面图

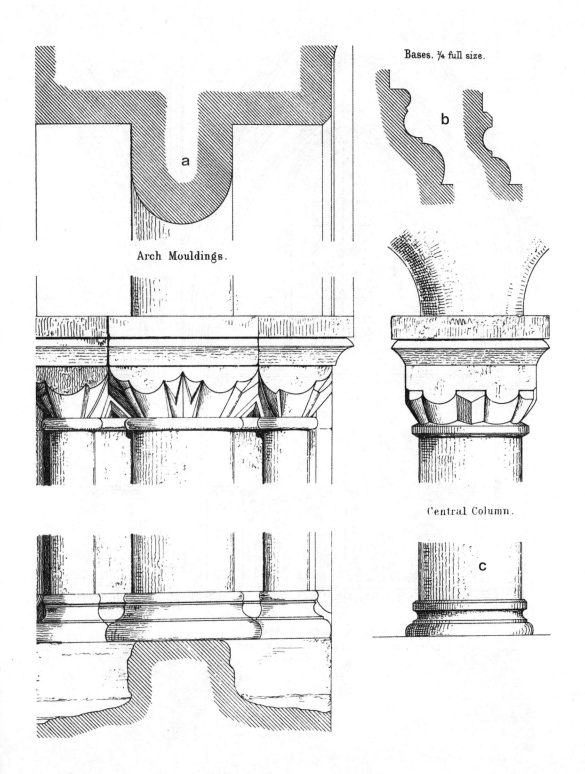

萨塞克斯郡伯克斯格罗夫教堂回廊细部：a. 拱线脚　b. 柱础　c. 中心柱

NEW SHOREHAM CHURCH, SUSSEX.
TRIFORIUM & CLERESTORY, NORTH SIDE
OF CHANCEL.

萨塞克斯郡新肖勒姆教堂圣坛北侧拱廊和高窗

Moulding B.

Corbels at A.

The letters refer to pl. 31.

Plan and details of Central pier, C.

Base D.

Section of Corbel at ✚

Central Shaft of the next compartment to the one given.

萨塞克斯郡新肖勒姆教堂：a.A 处的梁托 b.B 处的线脚 c.中心集束柱平面及细部 d.D 处的柱础 e.梁托的剖面 f.从特定一间到另一间的中心柱子

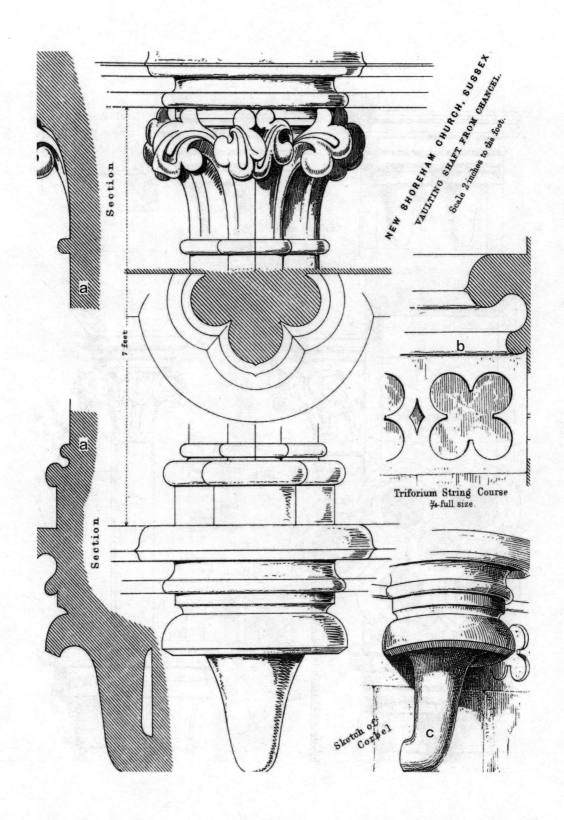

萨塞克斯郡新肖勒姆教堂圣坛中承肋柱：a. 剖面图　b. 拱廊束带层　c. 梁托素描

萨塞克斯郡伯克斯格罗夫教堂：a. 唱诗室一间（内部）　b. 剖面

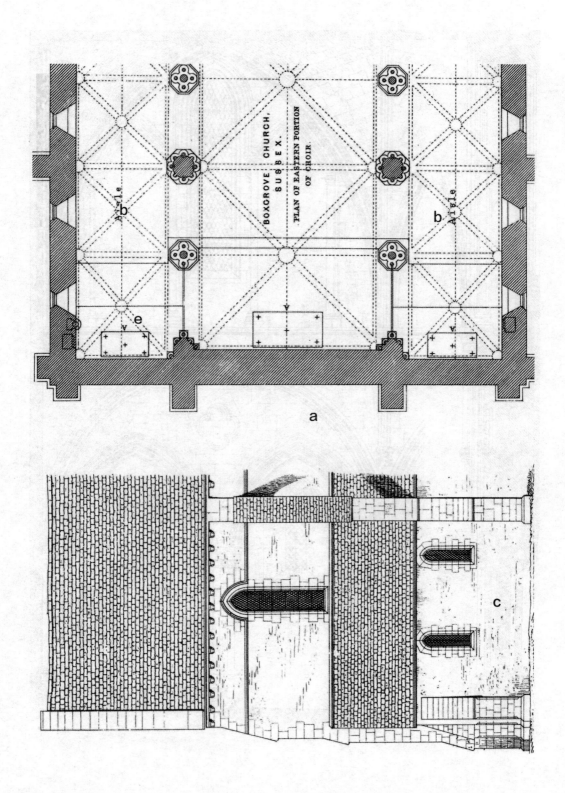

萨塞克斯郡伯克斯格罗夫教堂：a. 唱诗室东侧局部平面　b. 侧廊　c. 唱诗室一间（外部）　e. A 为古时圣餐台位置

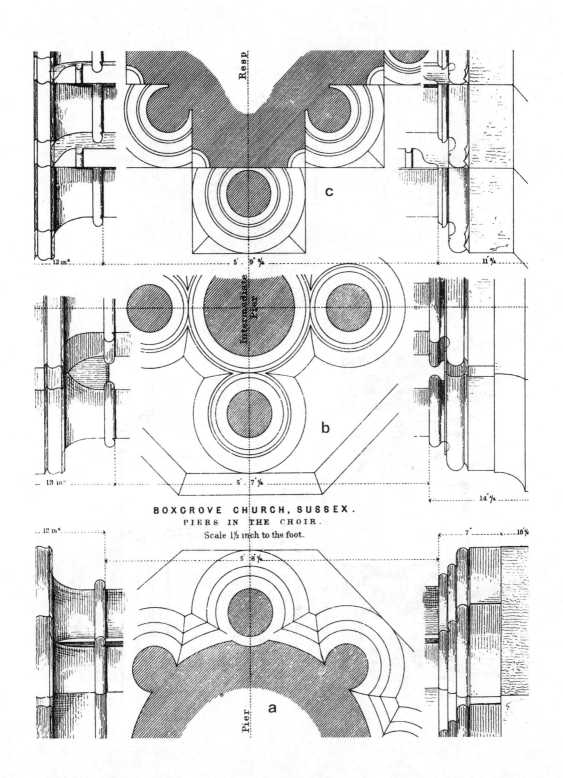

Resp.

Intermediate Pier

BOXGROVE CHURCH, SUSSEX.
PIERS IN THE CHOIR.
Scale 1½ inch to the foot.

Pier

萨塞克斯郡伯克斯格罗夫教堂唱诗室内集束柱：a. 主要集束柱　b. 次要集束柱　c. 壁联

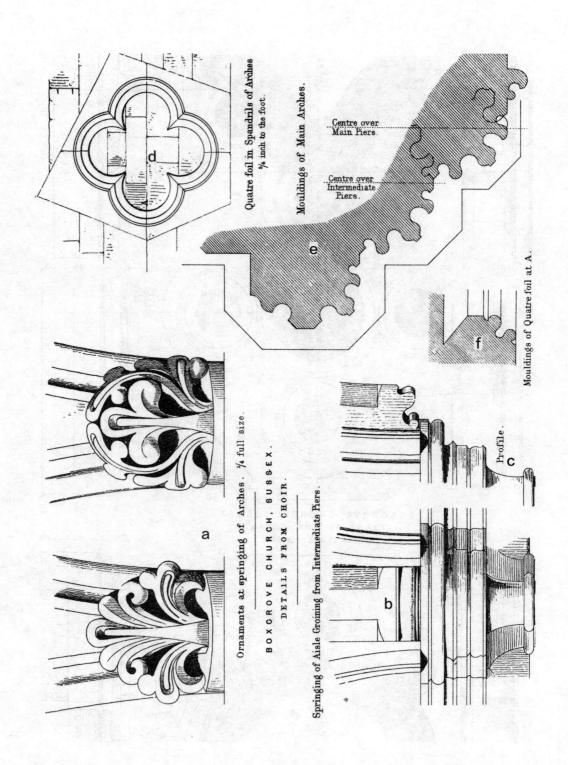

Quatre foil in Spandrils of Arches.
¾ inch to the foot.

Mouldings of Main Arches.

Centre over Main Piers.

Centre over Intermediate Piers.

Mouldings of Quatre foil at A.

Profile.

Ornaments at springing of Arches. ¾ full size.

BOXGROVE CHURCH, SUSSEX.
DETAILS FROM CHOIR.

Springing of Aisle Groining from Intermediate Piers.

萨塞克斯郡伯克斯格罗夫教堂唱诗室内细部：a.拱起拱点处装饰　　b.侧廊拱顶中墩上起拱处　　c.轮廓　　d.拱肩上四叶饰　　e.主要拱线脚　　f.A处四叶饰线脚

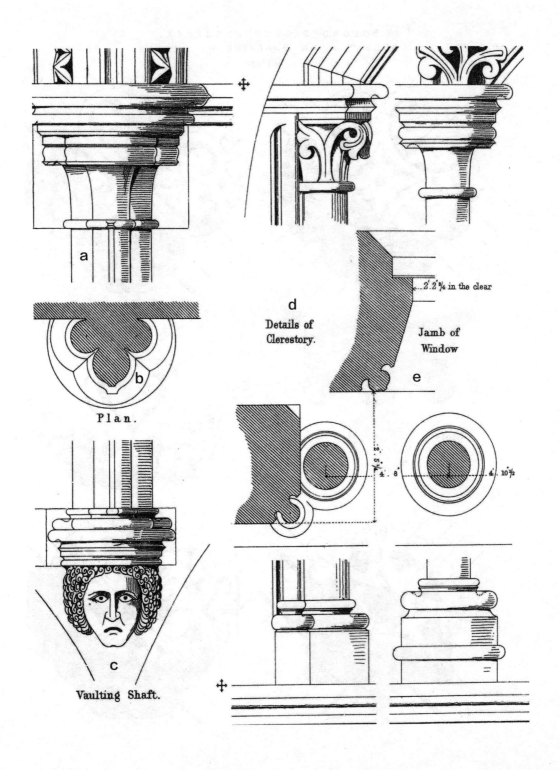

a

b

Plan.

c

Vaulting Shaft.

d

Details of
Clerestory.

2.2 ¾ in the clear

Jamb of
Window

e

萨塞克斯郡伯克斯格罗夫教堂唱诗室内部：a. 波贝克大理石　b. 平面　c. 承肋柱　d. 高窗细部　e. 窗户侧壁

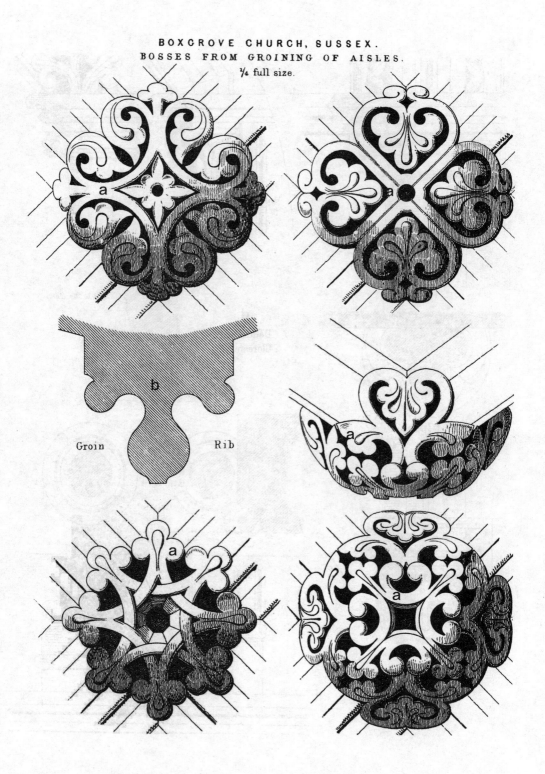

BOXGROVE CHURCH, SUSSEX.
BOSSES FROM GROINING OF AISLES.
¼ full size.

Groin        Rib

萨塞克斯郡伯克斯格罗夫教堂：a. 侧廊交叉拱上凸饰　　b. 穹棱

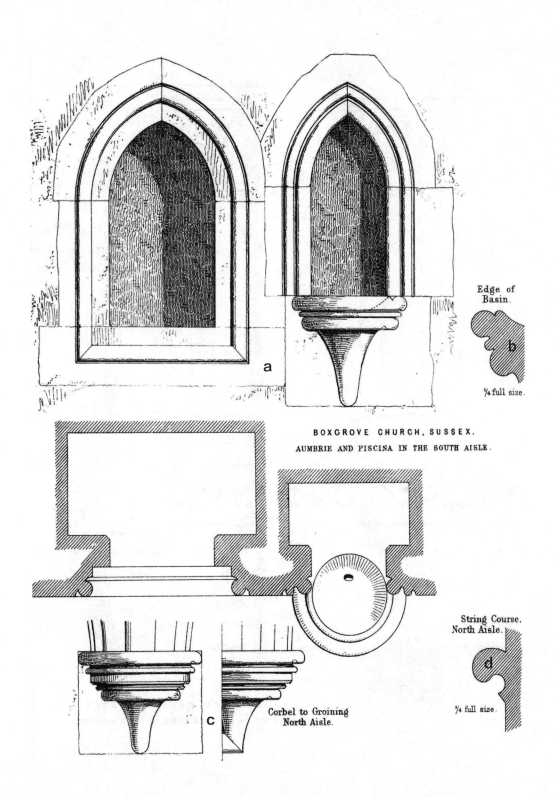

Edge of
Basin.

b

3/4 full size.

BOXCROVE CHURCH, SUSSEX.
AUMBRIE AND PISCINA IN THE SOUTH AISLE.

a

Corbel to Groining
North Aisle.

c

String Course.
North Aisle.

d

3/4 full size.

萨塞克斯郡伯克斯格罗夫教堂：a. 南侧廊中圣器壁龛和排水石盆　b. 石盆边缘　c. 北侧廊连接交叉拱梁托　d. 北侧廊束带层

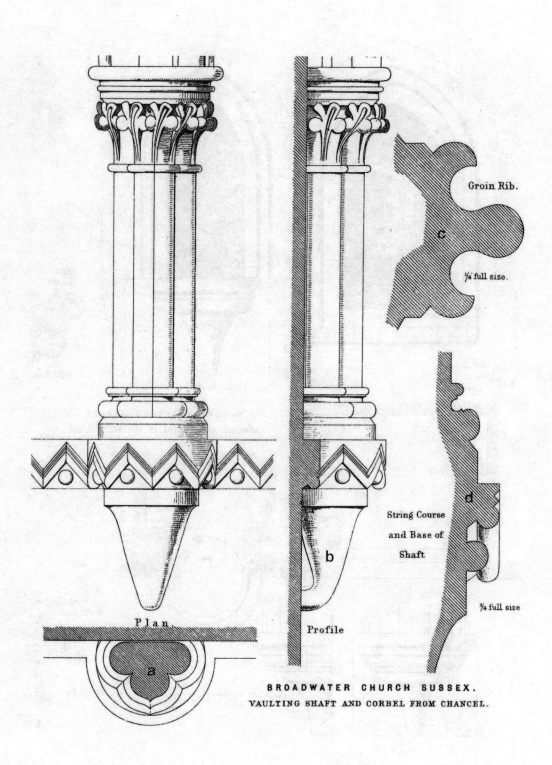

Groin Rib.

¼ full size.

String Course
and Base of
Shaft

¼ full size

BROADWATER CHURCH SUSSEX.
VAULTING SHAFT AND CORBEL FROM CHANCEL.

萨塞克斯郡布罗德沃特尔教堂圣坛内承肋柱和梁托：a. 平面图　b. 侧面图　c. 穹棱　d. 柱子束带层和柱础

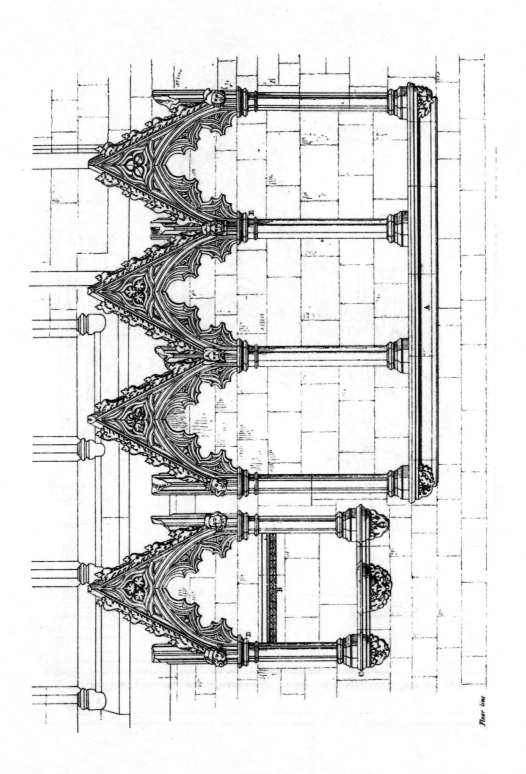

萨塞克斯郡温切尔西教堂南侧廊排水石盆和牧师席

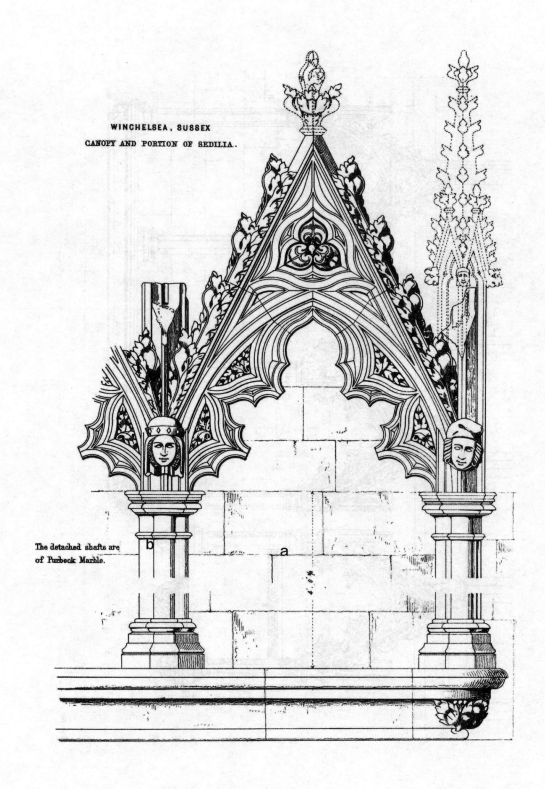

WINCHELSEA, SUSSEX
CANOPY AND PORTION OF SEDILIA.

The detached shafts are
of Purbeck Marble.

萨塞克斯郡温切尔西教堂：a. 牧师席顶部及局部　b. 分离柱柱身为波贝克大理石

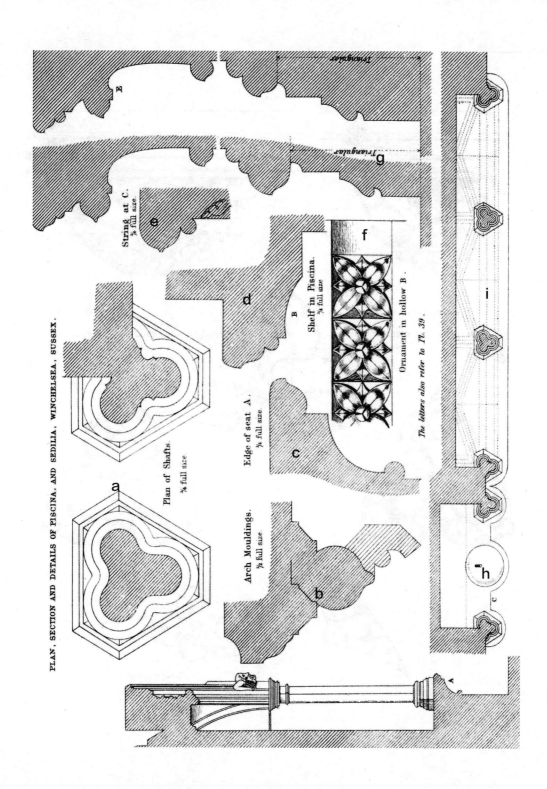

萨塞克斯郡温切尔西教堂排水石盆、牧师席平面剖面及细部：a. 柱子平面　b. 拱线脚　c. 座椅 A 的边缘　d. 排水石盆内搁板　e. C 处的束带　f. B 处的凹槽处装饰　g. 三角形　h. 排水石盆　i. 牧师席

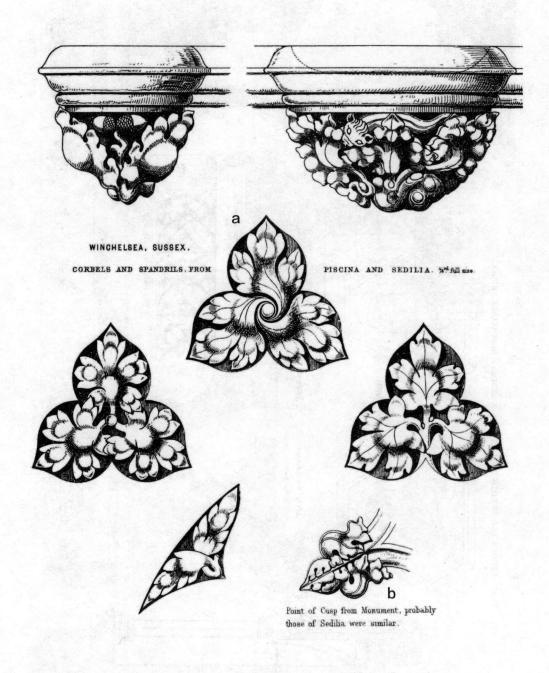

WINCHELSEA, SUSSEX.

CORBELS AND SPANDRILS, FROM

PISCINA AND SEDILIA. ⅓rd full size.

Point of Cusp from Monument, probably those of Sedilia were similar.

萨塞克斯郡温切尔西教堂：a. 排水石盆、牧师席上的梁托和菱形装饰　b. 遗迹中的尖头装饰或许牧师席中有类似装饰

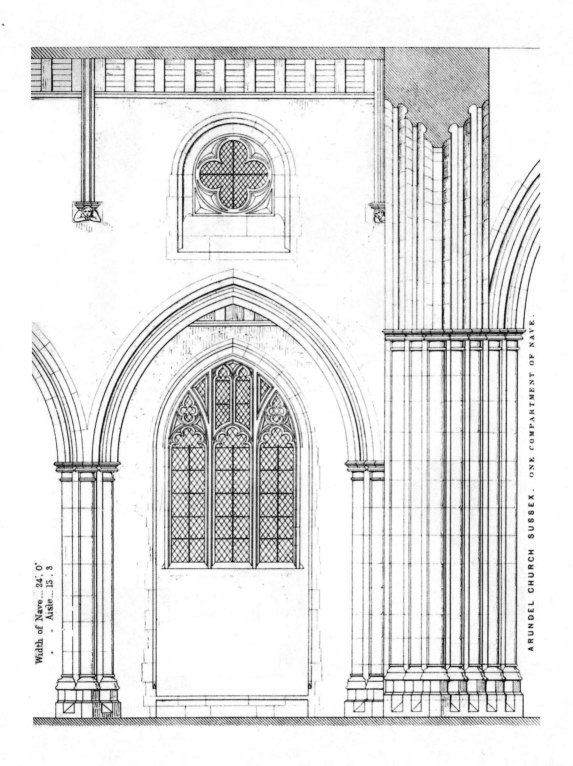

Width of Nave...24ʹ,0ʺ
" Aisle...15ʹ,3ʺ

ARUNDEL CHURCH SUSSEX. ONE COMPARTMENT OF NAVE.

萨塞克斯郡阿伦德尔教堂一间中殿

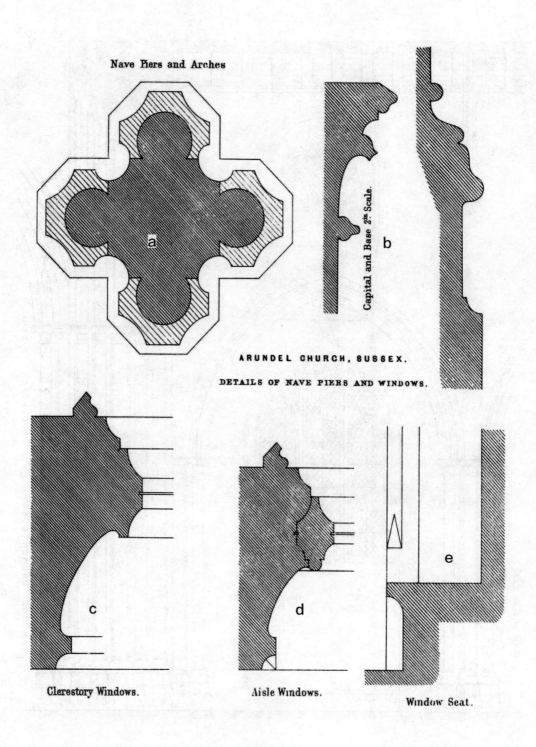

Nave Piers and Arches

a

Capital and Base 2ⁿ Scale.

b

ARUNDEL CHURCH, SUSSEX.

DETAILS OF NAVE PIERS AND WINDOWS.

c

d

e

Clerestory Windows.

Aisle Windows.

Window Seat.

萨塞克斯郡阿伦德尔教堂中殿集束柱和窗户细部：a. 中殿集束柱和拱　b. 柱头和柱础　c. 高窗　d. 侧廊窗户
e. 靠窗座椅

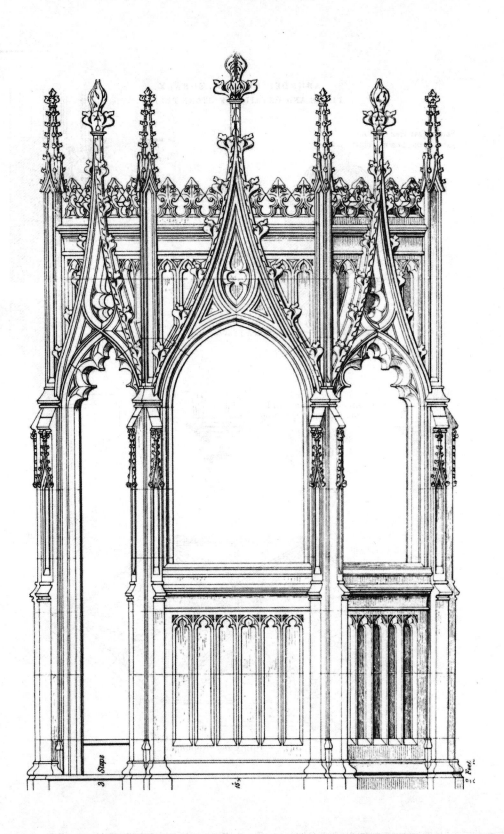

萨塞克斯郡阿伦德尔教堂中殿石制布道坛

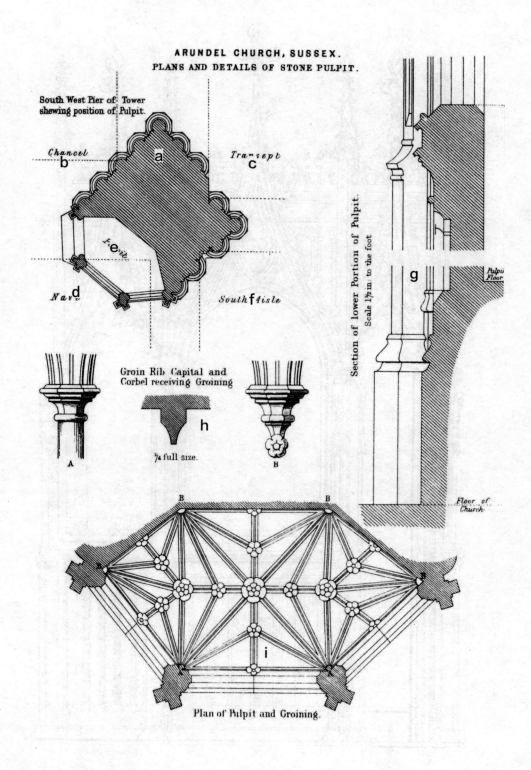

ARUNDEL CHURCH, SUSSEX.
PLANS AND DETAILS OF STONE PULPIT.

South West Pier of Tower
shewing position of Pulpit.

*Chancel*

*Transept*

*Nave*

*South Aisle*

Section of lower Portion of Pulpit.
Scale 1½ in. to the foot

Pulpit Floor

Groin Rib Capital and
Corbel receiving Groining

¾ full size.

Floor of Church

Plan of Pulpit and Groining.

萨塞克斯郡阿伦德尔教堂石制布道坛平面及细部：a. 展示布道坛位置的塔楼西南侧集束柱　b. 圣坛　c. 十字翼部
d. 中殿　e. 布道坛　f. 南侧廊　g. 布道坛低处剖面　h. 穹棱、柱头及承载穹顶的梁托　i. 布道坛及穹顶平面

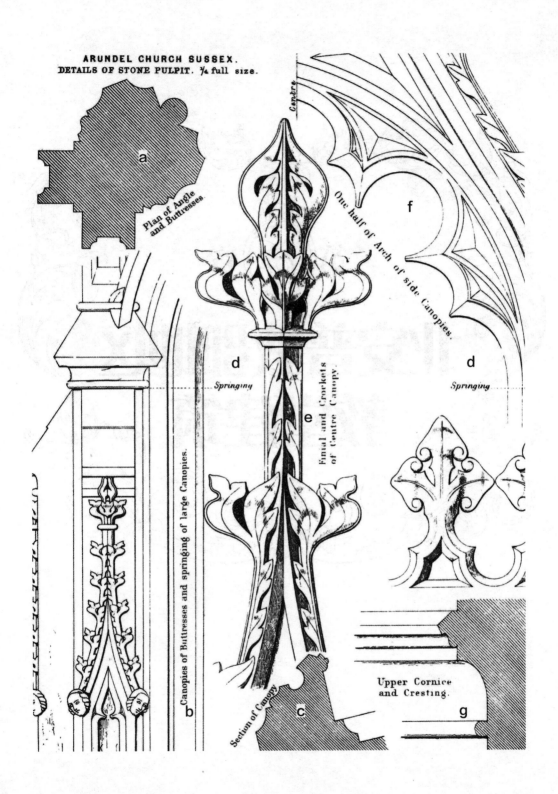

ARUNDEL CHURCH SUSSEX.
DETAILS OF STONE PULPIT. ¼ full size.

萨塞克斯郡阿伦德尔教堂石制布道坛细部：a. 转角平面　b. 扶壁顶部及大顶起拱面　c. 顶部剖面　d. 起拱面
e. 中心顶部尖形饰及卷花叶饰　f. 侧顶部拱一半　g. 上部檐口和顶饰

第 **6** 章

# 北安普敦郡地区教堂建筑

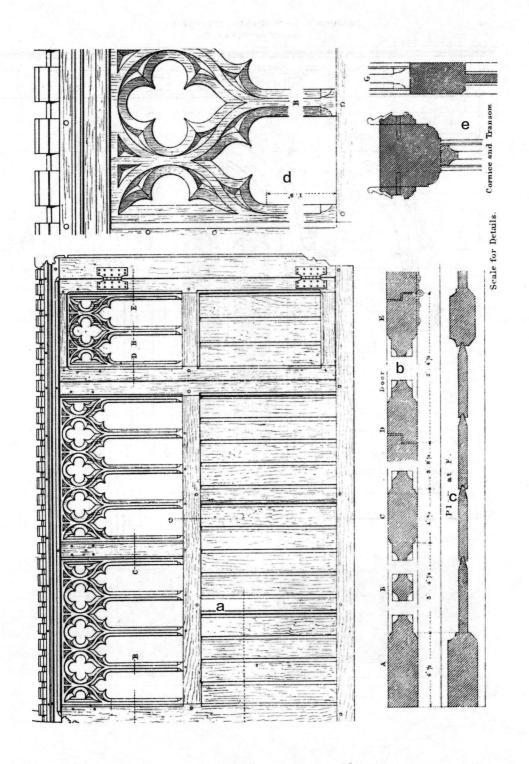

北安普敦郡盖丁顿教堂：a. 圣坛南侧隔断　b. 门　c. F 处的平面　d. 窗饰详细局部　e. 檐口和横楣

北安普敦郡邓福德教堂塔楼钟室：a. 窗户内部拱　b. 平面

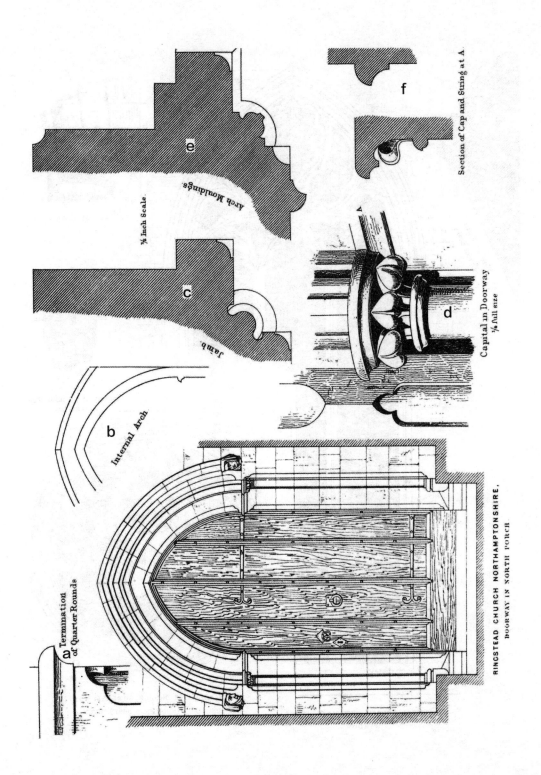

北安普敦灵斯特德教堂北门廊大门：a. 四分圆线脚端部　　b. 内侧拱　　c. 侧壁　　d. 门上柱头　　e. 拱线脚　　f. A 处的柱头和束带剖面

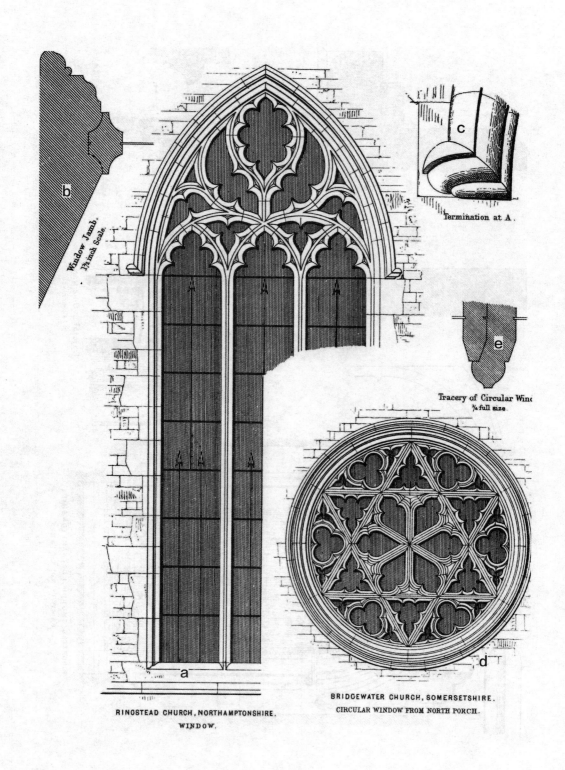

北安普敦灵斯特德教堂：a. 窗户　b. 窗户侧壁　c. A处的端头
索美赛特夏郡布里奇沃特教堂：d. 北门廊圆窗　e. 圆窗窗饰

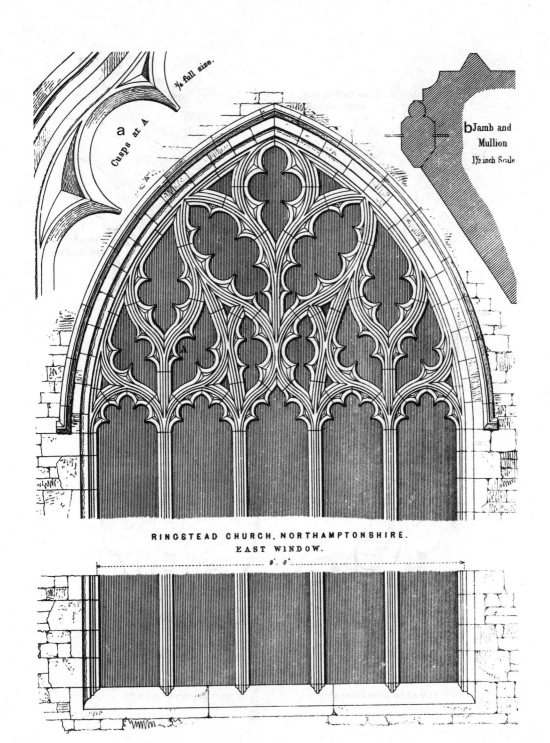

北安普敦灵斯特德教堂东窗：a.A处的尖头装饰　b.侧壁和竖框

STONE CROSS, QUARTER FULL SIZE, FROM A FRAGMENT
FOUND IN HICHAM FERRERS CHURCH, NORTHANTS.

a. 北安普敦郡海厄姆费雷教堂发现碎片中的石制十字架　　b. 来自索美赛特夏郡特鲁尔教堂　　c. 来自林肯郡弗莱斯顿教堂

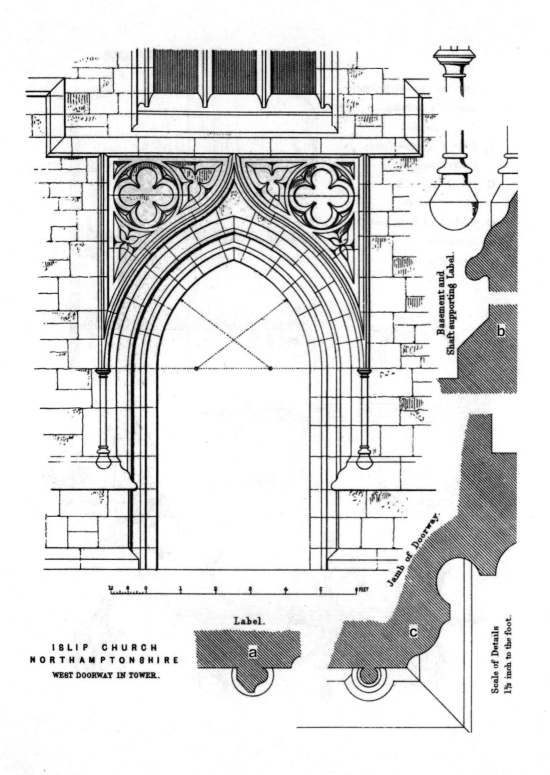

ISLIP CHURCH
NORTHAMPTONSHIRE
WEST DOORWAY IN TOWER.

Label.

Basement and Shaft supporting Label.

Jamb of Doorway.

Scale of Details
1⅓ inch to the foot.

北安普敦郡艾斯利普教堂塔楼西门：a. 披水石　　b. 底座及支撑披水石的柱子　　c. 大门侧壁

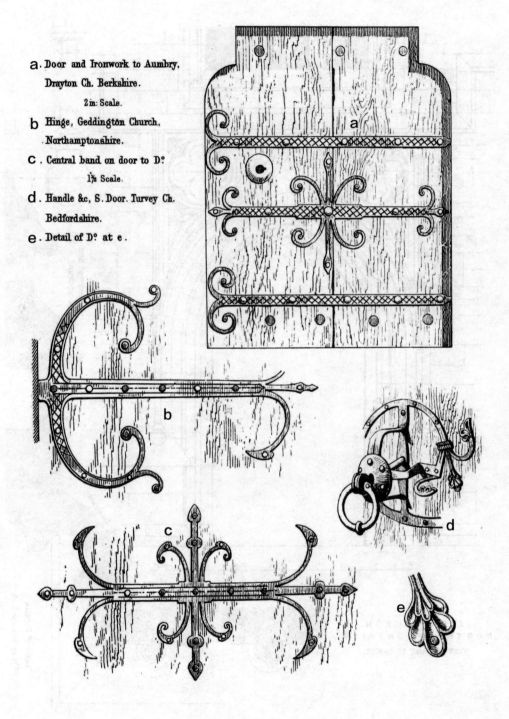

a. Door and Ironwork to Aumbry, Drayton Ch. Berkshire.
2 in: Scale.

b Hinge, Geddington Church, Northamptonshire.

c. Central band on door to D°
1½ Scale.

d. Handle &c, S. Door. Turvey Ch. Bedfordshire.

e. Detail of D° at e.

a. 伯克郡德雷顿教堂圣器壁龛门　　b. 北安普敦郡盖丁顿教堂铰链　　c.d 处门上的中心带　　d. 贝德弗德郡特威教堂门上把手　　e. 门上 e 处的细部

第 7 章

# 沃里克郡地区教堂建筑

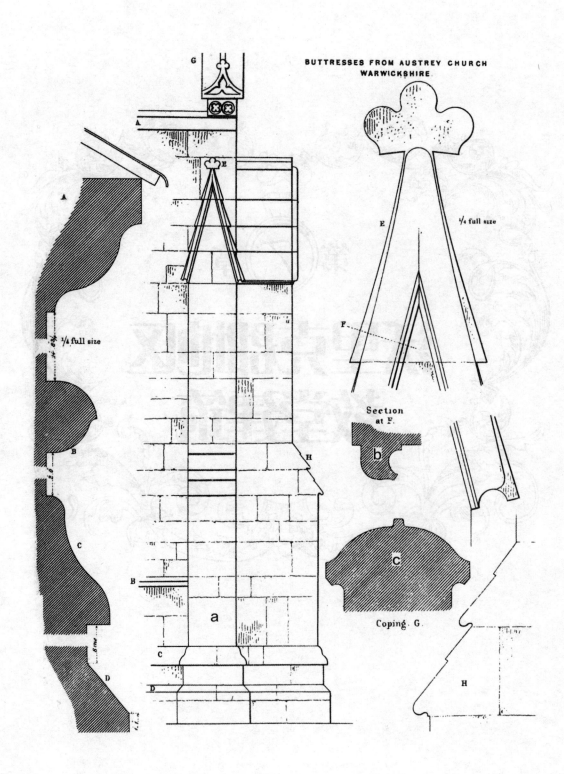

BUTTRESSES FROM AUSTREY CHURCH WARWICKSHIRE.

沃里克郡奥斯特利教堂：a.扶壁　b.F 处的剖面　c.G 处的压顶

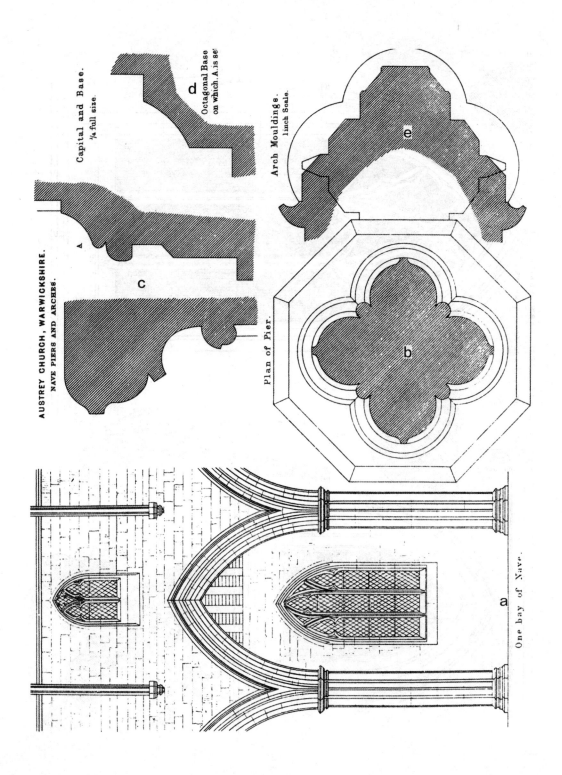

沃里克郡奥斯特利教堂中殿集束柱和拱：a. 中殿一跨　　b. 集束柱平面　　c. 柱头和柱础　　d.A 所放置的八边形底座
e. 拱线脚

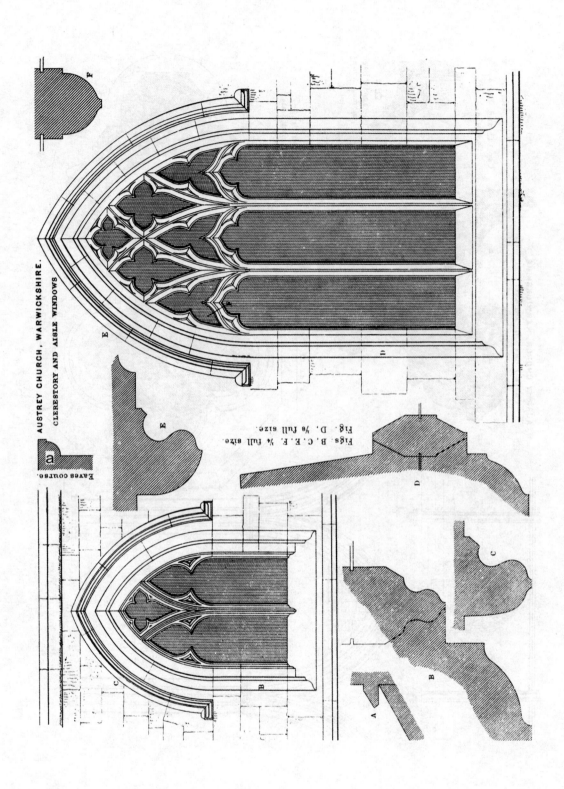

沃里克郡奥斯特利教堂高窗和走廊窗户：a. 檐层

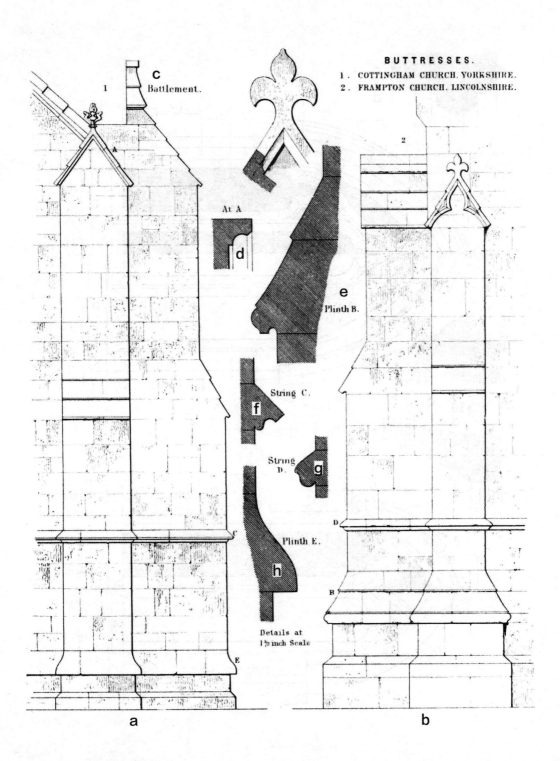

扶壁：a. 约克郡科廷厄姆教堂　　b. 林肯郡弗兰普顿教堂　　c. 雉堞墙（城垛）　　d.A 处　　e.b 处的底座　　f.C 处的束带　　g.D 处的束带　　h.E 处的底座

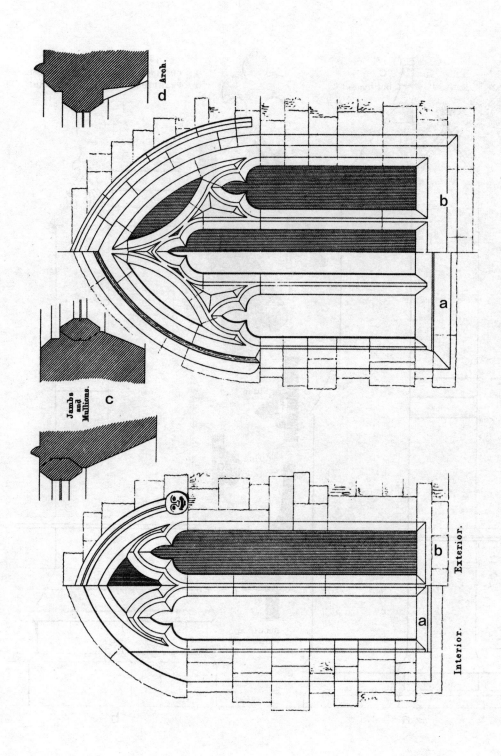

沃里克郡伯克斯韦尔教堂南侧走廊和圣坛窗户：a. 内侧　　b. 外侧　　c. 侧壁和竖框　　d. 拱

Exterior.   Interior.

Section.

Plan.

KENILWORTH CHURCH, WARWICKSHIRE.
COUPLED WINDOWS IN SOUTH AISLE.

Centre pier. 1½ inch Scale

沃里克郡凯尼尔沃思教堂：a. 南侧走廊双扇窗   b. 内侧   c. 外侧   d. 平面   e. 剖面   f. 中间束柱

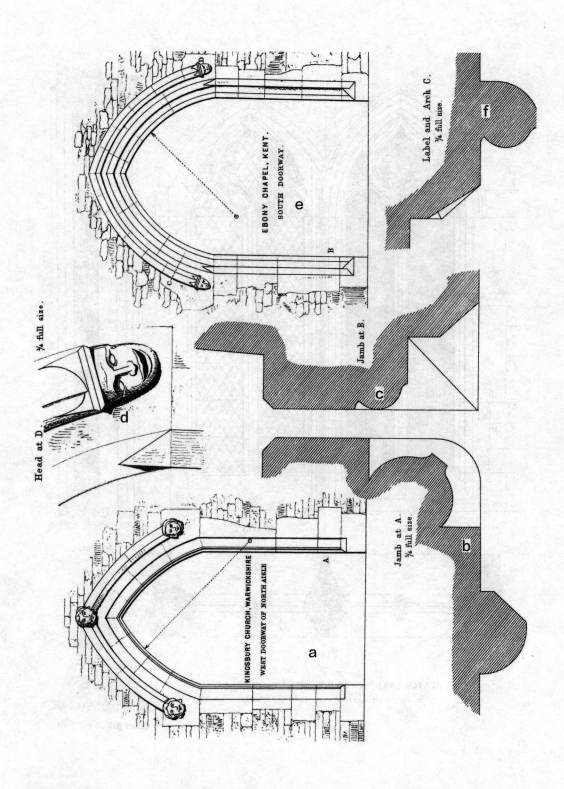

沃里克郡金斯伯里教堂：a. 北侧走廊西门　b.A 处的侧壁　c.B 处的侧壁　d.D 处的头像

肯特郡埃博尼小教堂：e. 南门　f.C 处披水石和拱剖面

第 8 章

# 萨福克郡地区教堂建筑

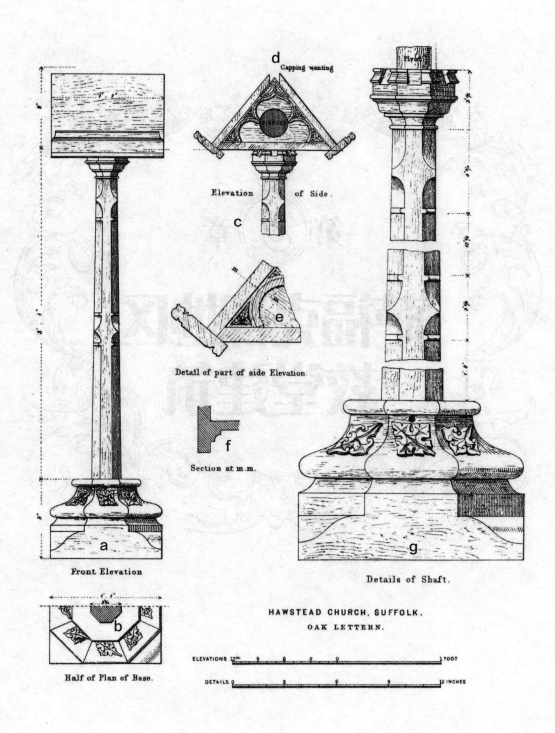

d Capping wanting

Elevation    of Side.

c

Front Elevation

m m

e

Detail of part of side Elevation.

f

Section at m.m.

Details of Shaft.

a

Front Elevation

b

Half of Plan of Base.

g

HAWSTEAD CHURCH, SUFFOLK.
OAK LETTERN.

ELEVATIONS 12in. 9    6    3    0                1 FOOT

DETAILS 0          3          6          9          12 INCHES

萨福克霍斯特德教堂橡木讲经台：a.正立面    b.底座平面一半    c.侧立面    d.欠缺的压顶    e.侧立面局部细节    f.m-m 处剖面    g.柱身细部

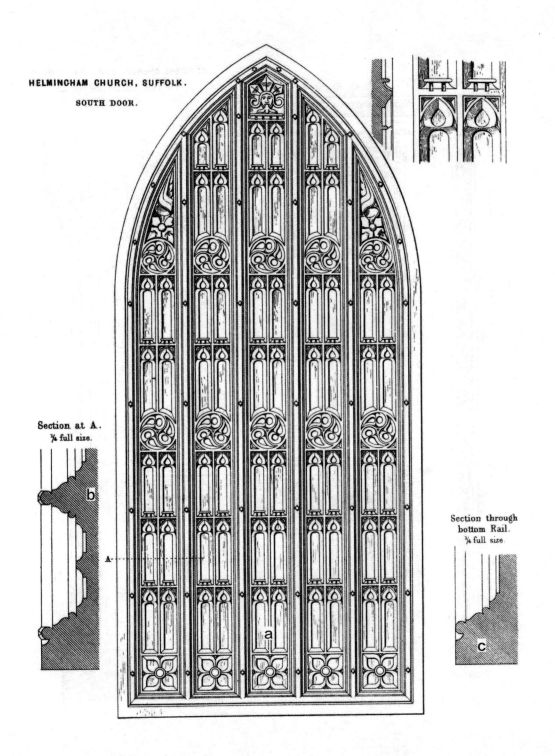

HELMINGHAM CHURCH, SUFFOLK.

SOUTH DOOR.

Section at A.
¾ full size.

Section through
bottom Rail.
¾ full size.

萨福克明海姆教堂：a.南门　b.A处的剖面　c.下横档剖面

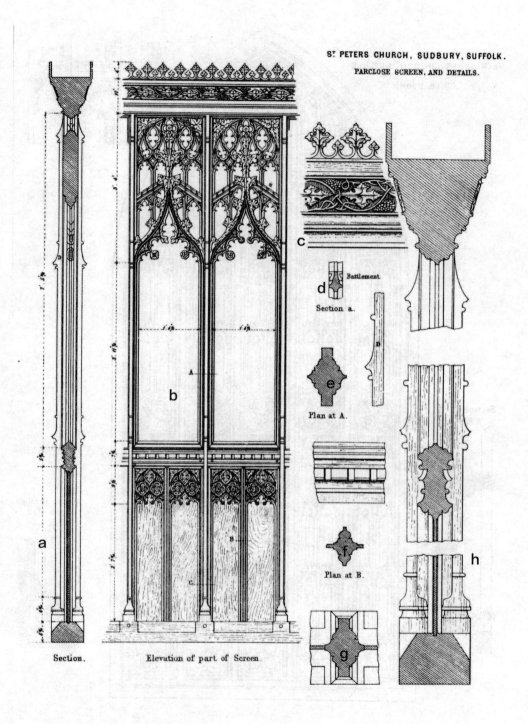

ST PETERS CHURCH, SUDBURY, SUFFOLK.

PARCLOSE SCREEN, AND DETAILS.

Battlement

Section a.

Plan at A.

Plan at B.

Section.

Elevation of part of Screen.

萨福克圣彼得教堂隔断及其细部：a. 剖面　b. 隔断局部立面　c. 城垛式装饰　d.a 处的剖面　e.A 处的平面　f.B 处的平面　g.C 处的详细平面　h. 详细的剖面

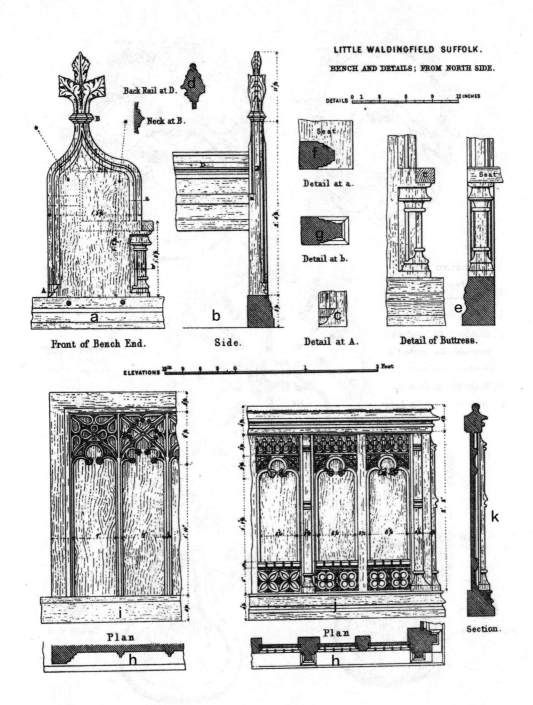

LITTLE WALDINGFIELD SUFFOLK.

BENCH AND DETAILS; FROM NORTH SIDE.

DETAILS 0 1 3 6 12 INCHES

Back Rail at D.

Neck at B.

Seat

Detail at a.

Detail at b.

Detail at A.

Front of Bench End.

Side.

Detail of Buttress.

Seat

Section.

ELEVATIONS 12in Feet

Plan

Plan

萨福克瓦尔丁菲尔德小教堂北侧长椅及细部：a. 长椅端部正面　b. 长椅端部侧面　c.A 处的细部　d.D 处的背部横木
e. 支撑细部　f.A 处的细部　g.B 处的细部　h. 平面长椅嵌板　i. 莱斯特教堂　j. 北安普敦郡盖丁顿教堂　k. 剖面

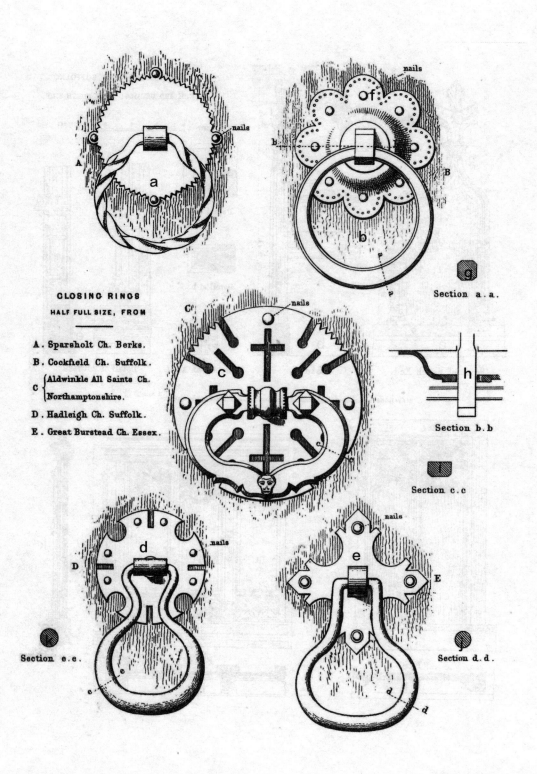

**CLOSING RINGS**
HALF FULL SIZE, FROM

A. Sparsholt Ch. Berks.
B. Cockfield Ch. Suffolk.
C. { Aldwinkle All Saints Ch. Northamptonshire.
D. Hadleigh Ch. Suffolk.
E. Great Burstead Ch. Essex.

门环：a. 伯克斯郡斯帕肖特教堂　b. 萨福克科克菲尔德教堂　c. 北安普敦郡埃尔德温科教堂　d. 萨福克哈德利教堂　e. 埃塞克斯博斯蒂德教堂　f. 钉子　g. 剖面a-a　h. 剖面b-b　i. 剖面c-c　j. 剖面d-d　k. 剖面e-e

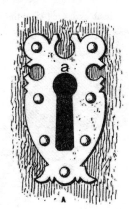

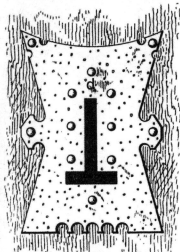

Key Plates; ⅜ full size,
   from

A. Ringsfield Church,
           Suffolk.

B. Dunsfold Church,
           Surrey.

C. Drayton Church,
           Berkshire.

D. St Mary's Church,
           Leicester.

E. Pytchley Church,
           Northamptonshire

F. Thurnham Church,
           Kent.

G. Pytchley Church
           Northamptonshire.

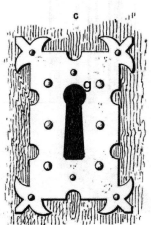

钥匙孔板：a.萨福克灵斯菲尔德教堂　b.萨里郡敦斯福德教堂　c.伯克郡德雷顿教堂　d.莱斯特圣玛丽教堂　e.北安普敦郡比其利教堂　f.肯特郡萨恩汉教堂　g.北安普敦郡比其利教堂

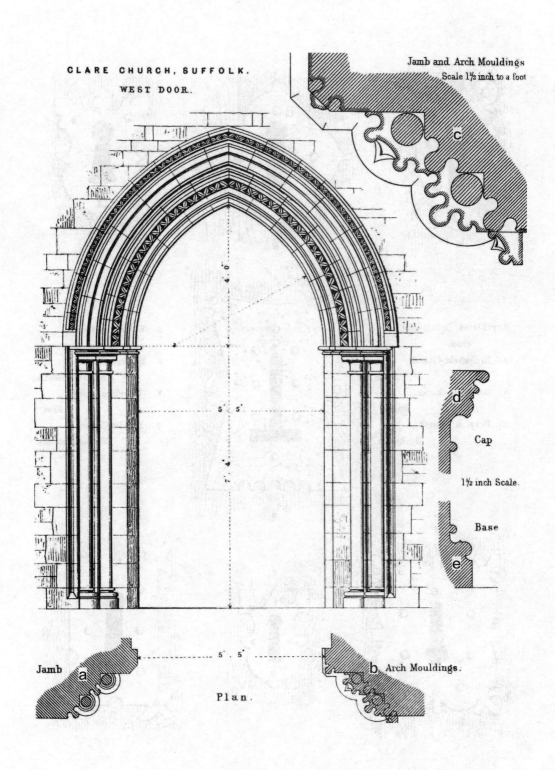

CLARE CHURCH, SUFFOLK.

WEST DOOR.

Jamb and Arch Mouldings
Scale 1½ inch to a foot

Cap

1½ inch Scale.

Base

Jamb

Plan.

Arch Mouldings.

萨福克郡克莱尔教堂西门：a. 侧壁　b. 拱线脚　c. 侧壁及拱线脚　d. 柱头　e. 柱础

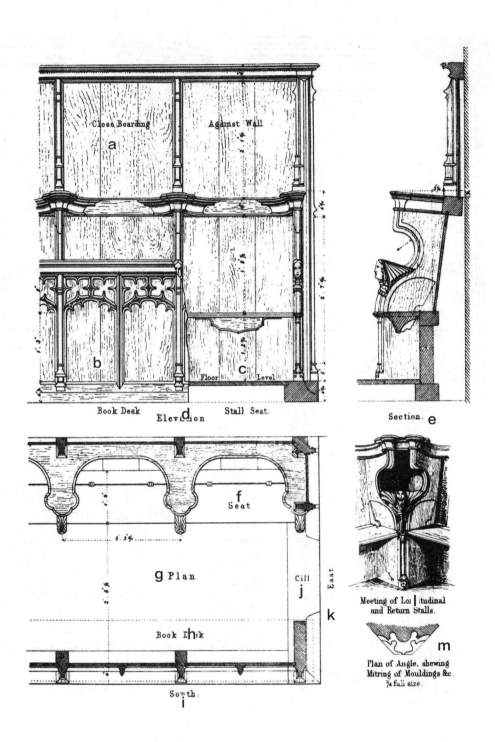

萨福克郡萨德伯里圣格里高利教堂 圣坛唱诗班座椅：a. 靠墙的封闭木板　b. 书桌　c. 唱诗班座椅　d. 立面　e. 剖面　f. 座位　g. 平面　h. 书桌　i. 南侧　j. 门槛　k. 东侧　l. 纵向座椅和转延座椅交汇处　m. 转角平面展现了线脚的斜接

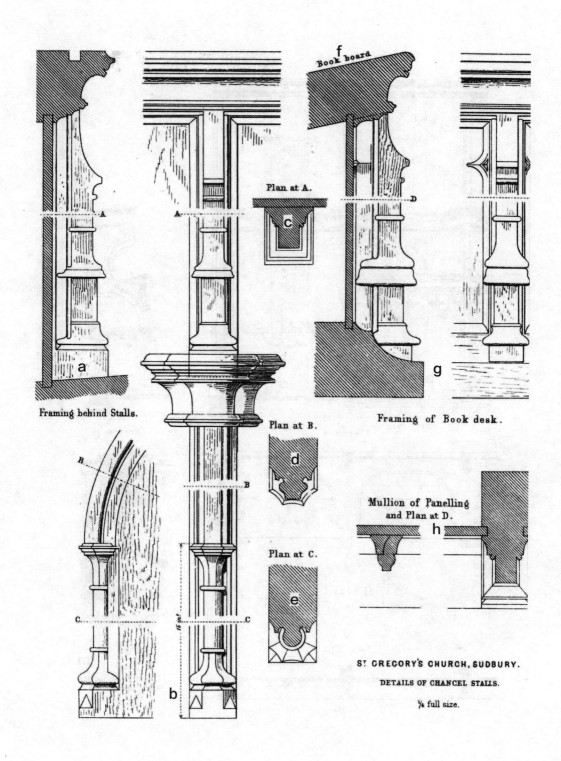

Framing behind Stalls.

Plan at A.

Book board

Plan at B.

Plan at C.

Framing of Book desk.

Mullion of Panelling
and Plan at D.

ST GREGORY'S CHURCH, SUDBURY.

DETAILS OF CHANCEL STALLS.

¼ full size.

萨福克郡萨德伯里圣格里高利教堂圣坛唱诗班座椅细部：a. 座椅后部框架　　b. 座椅扶手和框架　　c.A 处的平面
d.B 处的平面　　e.C 处的平面　　f. 置书板　　g. 书桌框架　　h. 嵌板竖框和 D 处的平面

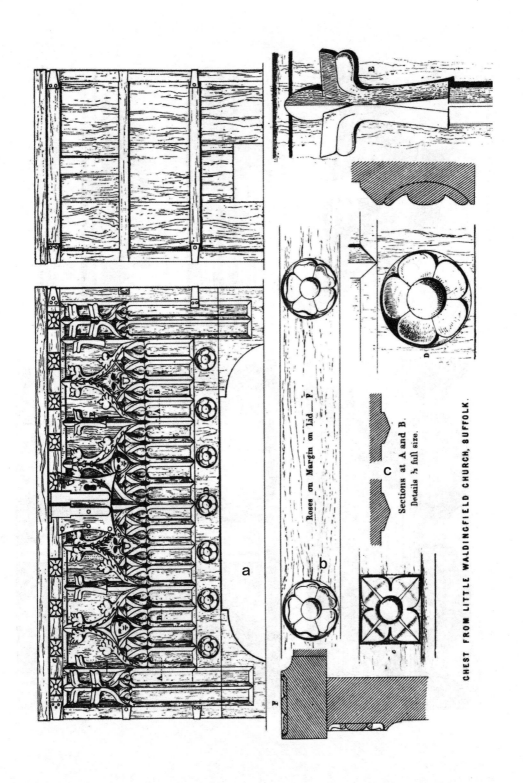

a. 萨福克郡瓦尔丁菲尔德小教堂里的箱子　　b.F 盖子边缘的蔷薇装饰　　c.A 和 B 处剖面

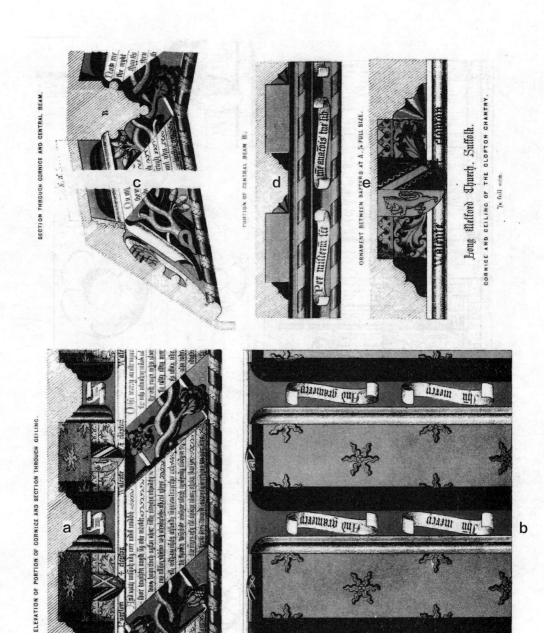

萨福克朗梅尔福德教堂克洛普顿礼拜堂檐口及天花：a. 檐口及立面及天花剖面局部　b. 天花局部平面　c. 檐口及环梁剖面　d. 环梁 B 的局部　e.A 处的椽子之间的装饰

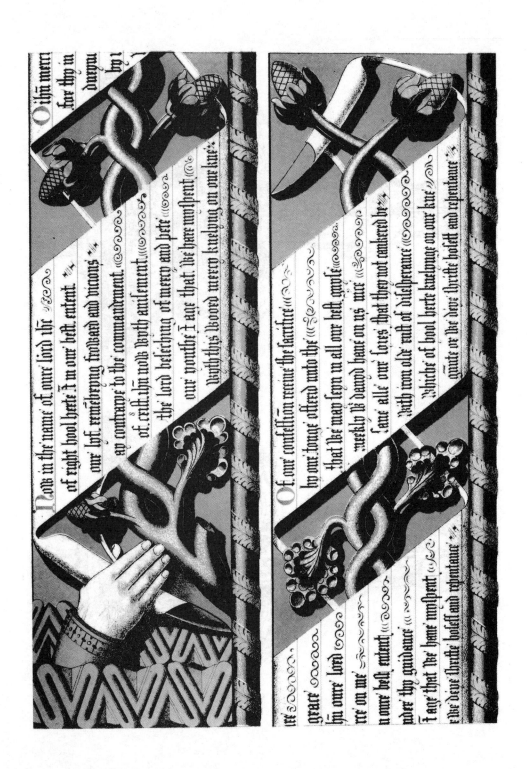

萨福克朗梅尔福德教堂克洛普顿礼拜堂檐口上画卷开端及结束位置

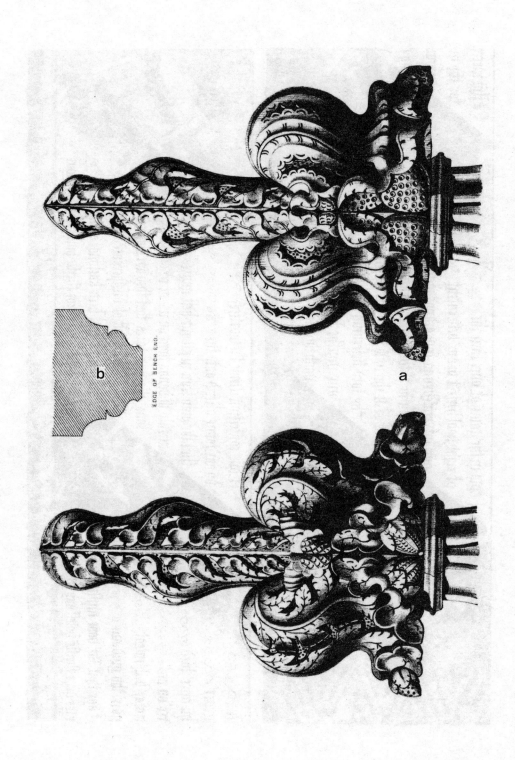

萨福克拉文纳姆教堂：a. 木制罂粟状装饰　　b. 长椅端头边缘

萨福克拉文纳姆教堂座椅上的橡木装饰

萨福克贝里圣埃德蒙兹圣玛丽教堂：南侧廊东部一间天花嵌板——原为约翰·巴雷特礼拜堂

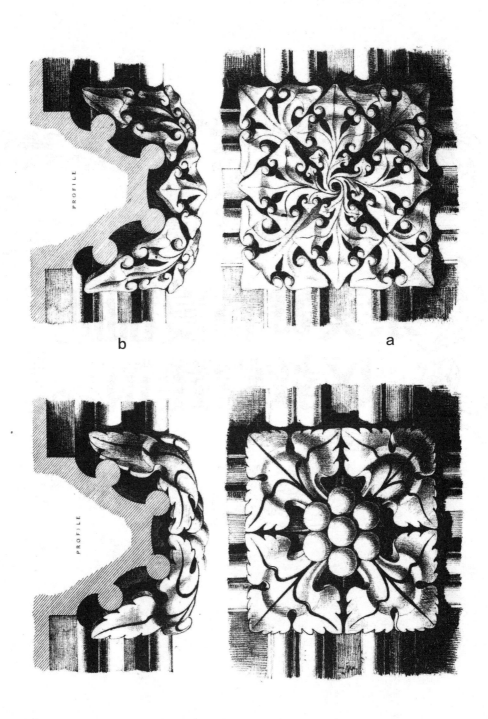

萨福克贝里圣埃德蒙兹圣玛丽教堂：a. 天花板上的橡木凸饰　b. 侧面

第9章

索美赛特夏郡地区教堂建筑

ST MARY'S CHURCH, BRIDGEWATER,
SOMERSETSHIRE.

DOORWAY OF NORTH PORCH.

索美赛特夏郡圣玛丽教堂北门廊大门

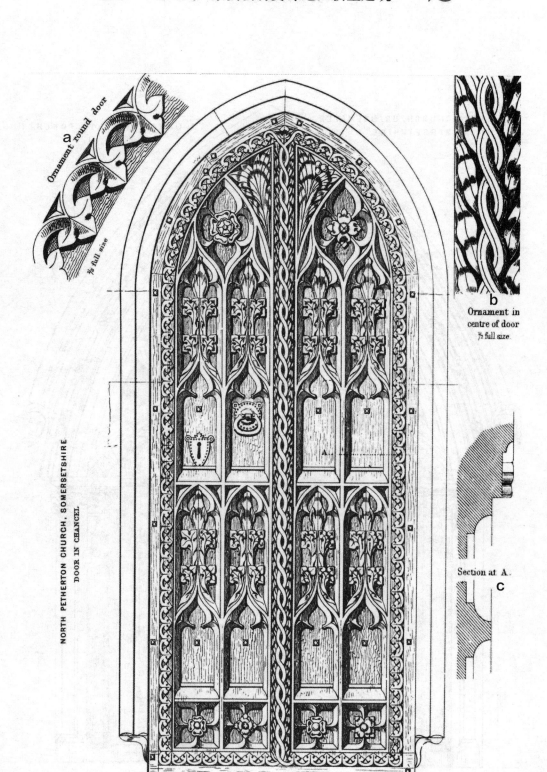

Ornament round door
½ full size
a

NORTH PETHERTON CHURCH, SOMERSETSHIRE
DOOR IN CHANCEL.

b
Ornament in centre of door
½ full size.

Section at A.
c

索美赛特夏郡北佩瑟顿教堂圣坛内大门：a. 环绕门的装饰　b. 门中心装饰　c.A 处的剖面

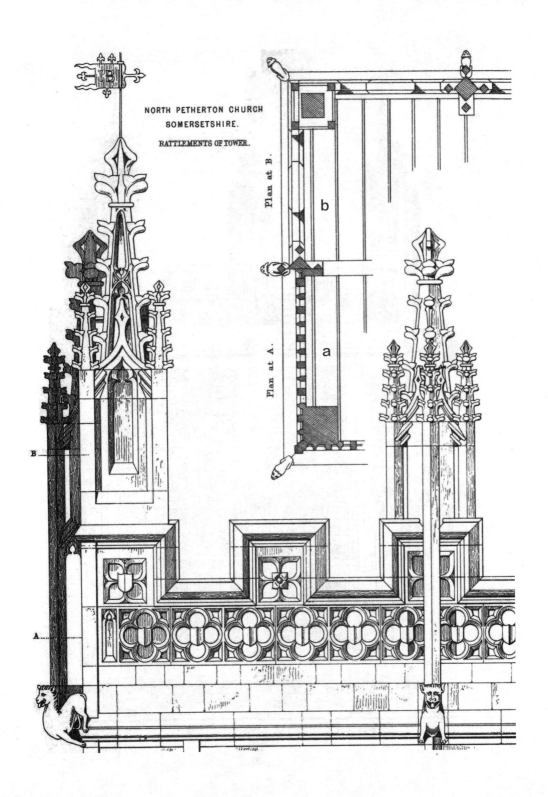

NORTH PETHERTON CHURCH
SOMERSETSHIRE.

BATTLEMENTS OF TOWER.

Plan at B.

Plan at A.

索美赛特夏郡北佩瑟顿教堂塔楼墙垛：a.A 处的平面　　b.B 处的平面

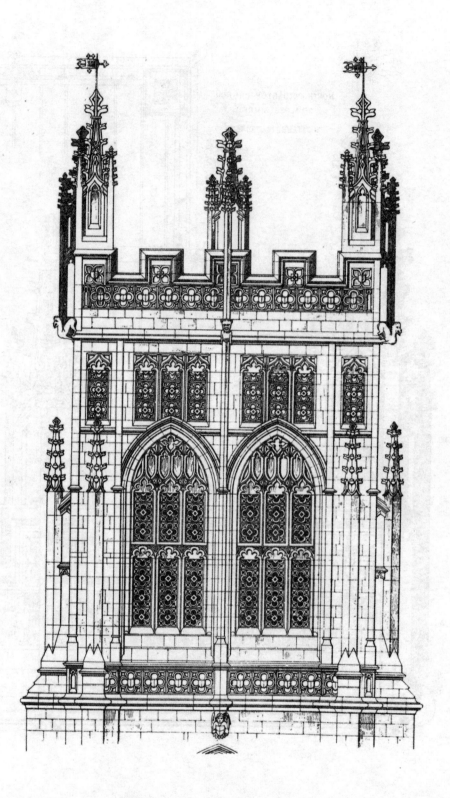

索美赛特夏郡北佩瑟顿教堂塔楼上部

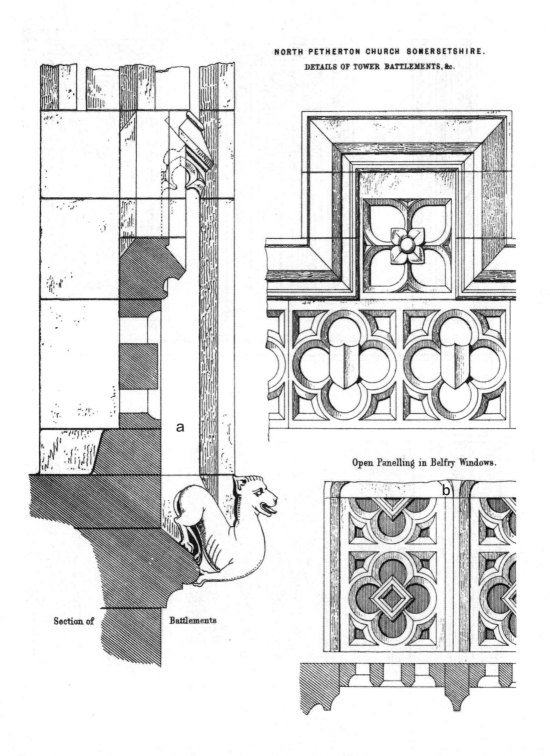

NORTH PETHERTON CHURCH SOMERSETSHIRE.
DETAILS OF TOWER BATTLEMENTS, &c.

Open Panelling in Belfry Windows.

Section of　　　Battlements

索美赛特夏郡北佩瑟顿教堂塔楼墙垛细部：a. 墙垛剖面　　b. 钟楼窗户可开启嵌板

TRULL CHURCH SOMERSETSHIRE
BENCH ENDS.

Carved Panels in Bench Ends.

索美赛特夏郡特鲁尔教堂：a. 长椅末端　b. 长椅末端的雕刻嵌板

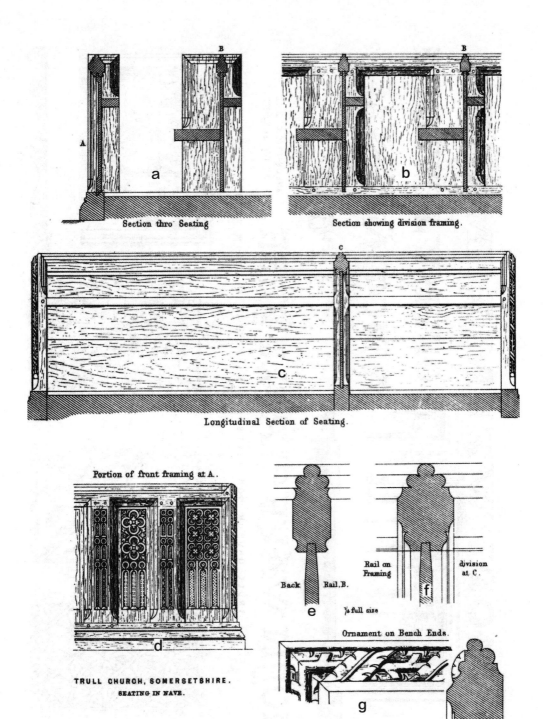

Section thro' Seating

Section showing division framing.

Longitudinal Section of Seating.

Portion of front framing at A.

Back Rail, B.

⅛ full size

Rail on Framing

division at C.

Ornament on Bench Ends.

TRULL CHURCH, SOMERSETSHIRE.

SEATING IN NAVE.

索美赛特夏郡特鲁尔教堂中殿座椅：a. 座椅剖面　b. 展示分隔框架的剖面　c. 座椅纵向剖面　d.A 处的正面框架的局部　e.B 处的靠背横木　f.C 处的分隔框架上横木　g. 长椅末端装饰

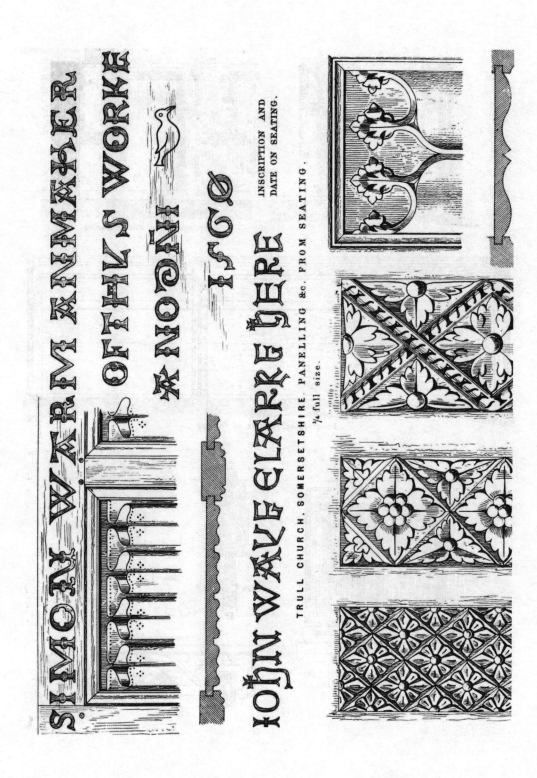

索美赛特夏郡特鲁尔教堂座椅嵌板座椅上铭文及日期座椅末端

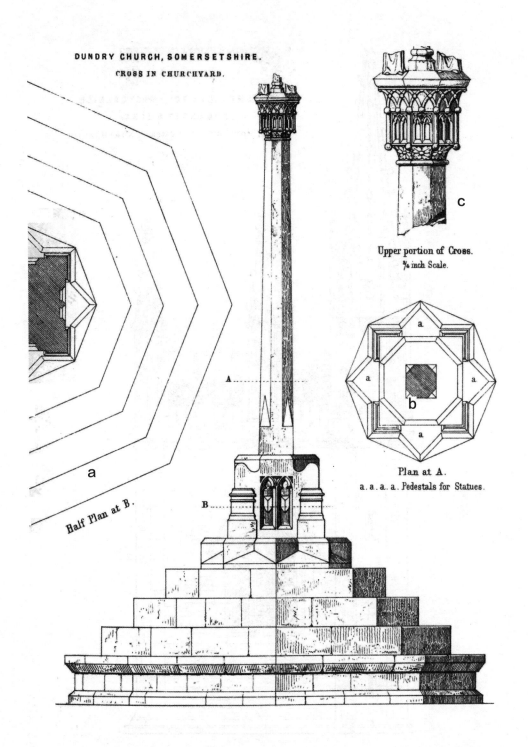

DUNDRY CHURCH, SOMERSETSHIRE.
CROSS IN CHURCHYARD.

Upper portion of Cross.
¾ inch Scale.

c

a
Half Plan at B.

A

B

a

a

b

a

a

Plan at A.
a. a. a. a. Pedestals for Statues.

索美赛特夏郡邓德里教堂教堂墓地内十字架：a.B 处的半平面　　b.A 处的平面　　c.十字架上部

ST JAMES CHURCH. WESTERLEIGH.
SOMERSETSHIRE.
UPPER PORTION OF TOWER—NORTH SIDE.

索美赛特夏郡圣詹姆斯教堂北侧塔楼上部

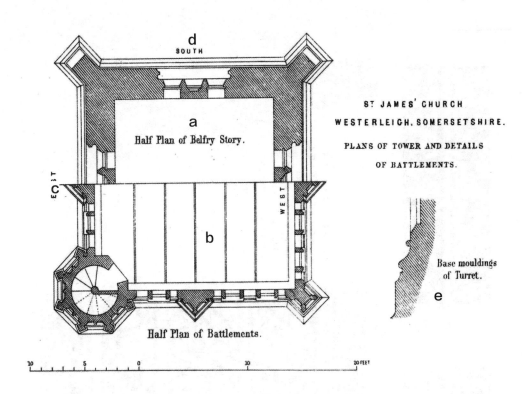

ST JAMES' CHURCH
WESTERLEIGH, SOMERSETSHIRE.
PLANS OF TOWER AND DETAILS
OF BATTLEMENTS.

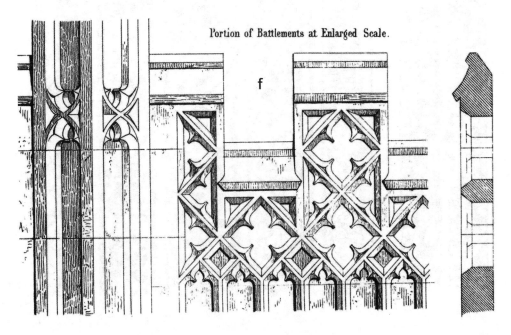

Portion of Battlements at Enlarged Scale.

索美赛特夏郡圣詹姆斯教堂塔楼平面及墙垛细部：a. 钟室层半平面　　b. 墙垛半平面　　c. 东侧　　d. 南侧　　e. 小塔底座线脚　　f. 墙垛局部

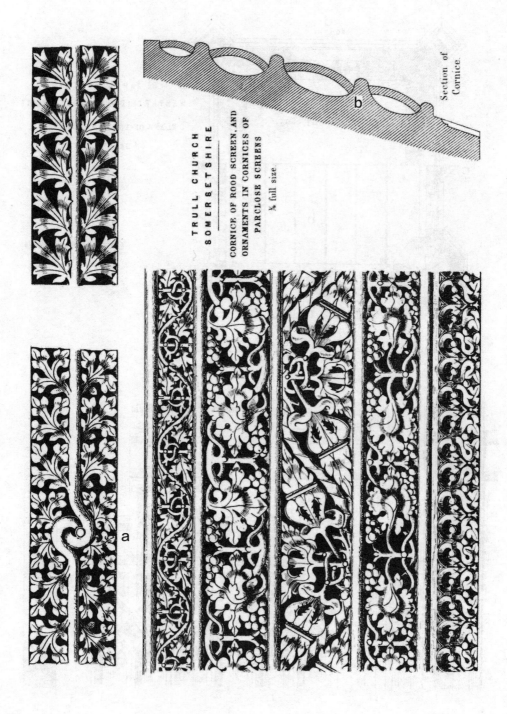

索美赛特夏郡特鲁尔教堂：a. 圣坛屏风檐板及分间隔屏檐板上装饰　　b. 檐板剖面

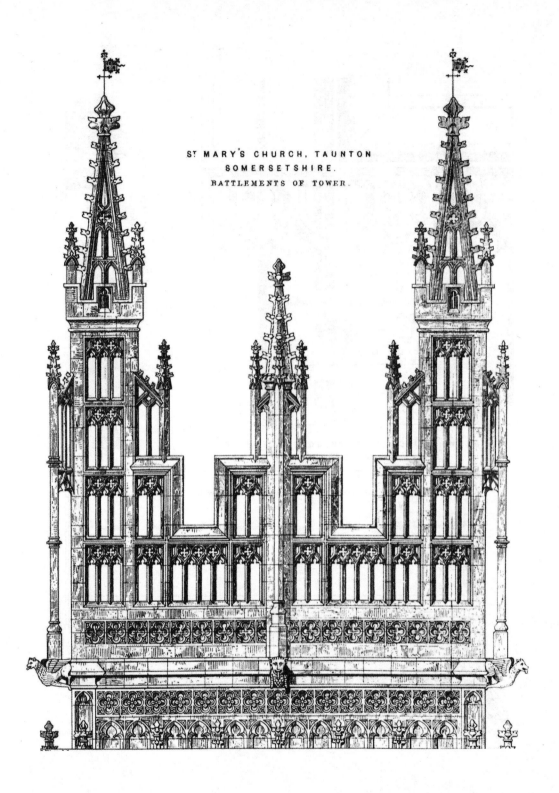

ST MARY'S CHURCH, TAUNTON
SOMERSETSHIRE.
BATTLEMENTS OF TOWER.

索美赛特夏郡陶顿圣玛丽教堂带小尖塔的墙垛

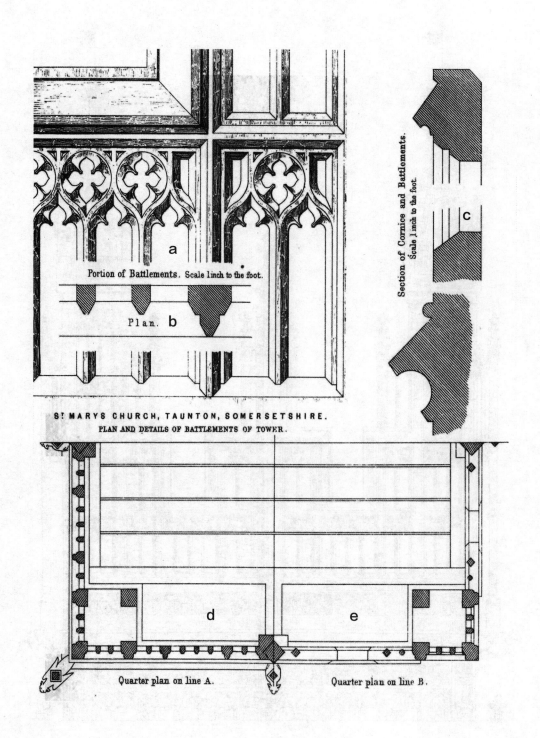

Portion of Battlements. Scale 1 inch to the foot.

Plan. b

Section of Cornice and Battlements.
Scale 1 inch to the foot.

S! MARYS CHURCH, TAUNTON, SOMERSETSHIRE.
PLAN AND DETAILS OF BATTLEMENTS OF TOWER.

Quarter plan on line A.　　　Quarter plan on line B.

索美赛特夏郡陶顿圣玛丽教堂带小尖塔的墙垛平面及细部：a. 墙垛局部　b. 平面　c. 檐口和墙垛剖面　d.A 处的剖四分之一平面　e.B 处的剖四分之一平面

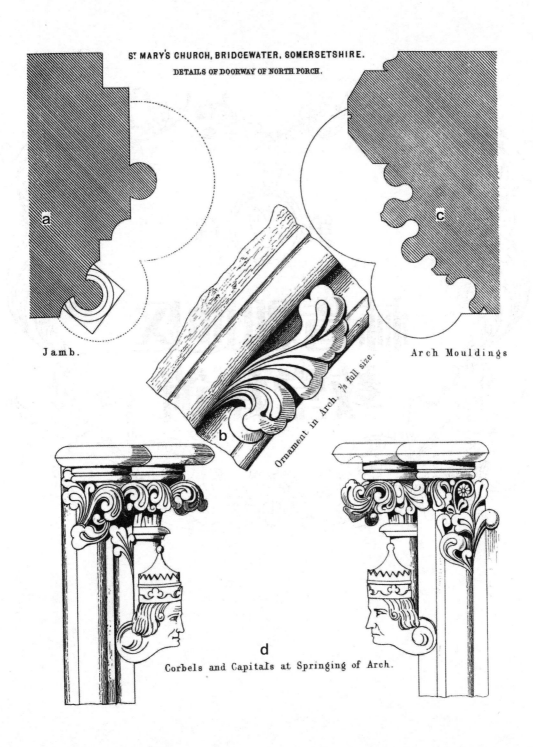

St. MARY'S CHURCH, BRIDGEWATER, SOMERSETSHIRE.
DETAILS OF DOORWAY OF NORTH PORCH.

Jamb.

Arch Mouldings

Ornament in Arch. ⅓ full size.

Corbels and Capitals at Springing of Arch.

索美赛特夏郡圣玛丽教堂北门廊大门细部：a. 侧壁　b. 拱上装饰　c. 拱线脚　d. 拱起的拱线处梁托和柱头

第 ⑩ 章

# 林肯郡地区教堂建筑

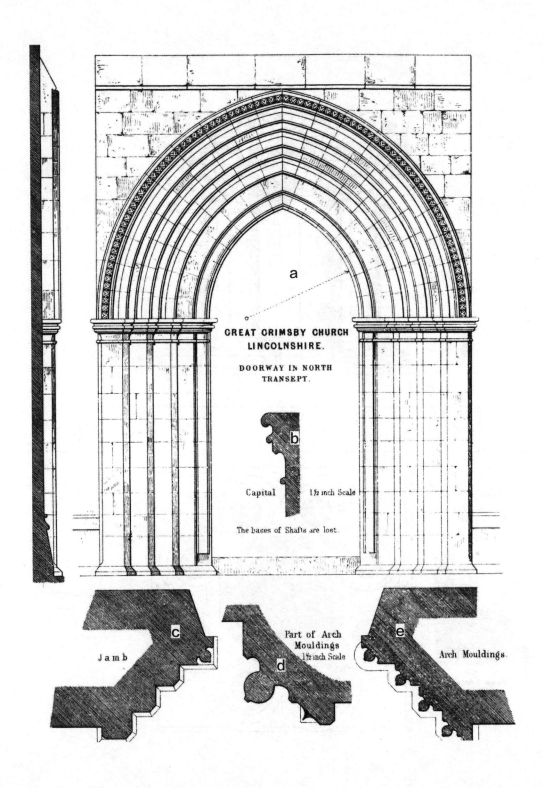

GREAT GRIMSBY CHURCH LINCOLNSHIRE.

DOORWAY IN NORTH TRANSEPT.

Capital      1½ inch Scale

The bases of Shafts are lost.

Part of Arch Mouldings 1½ inch Scale

Jamb

Arch Mouldings.

林肯郡格里姆斯比大教堂：a. 北耳堂大门　b. 柱头　c. 侧壁　d. 拱门装饰局部　e. 拱门线脚

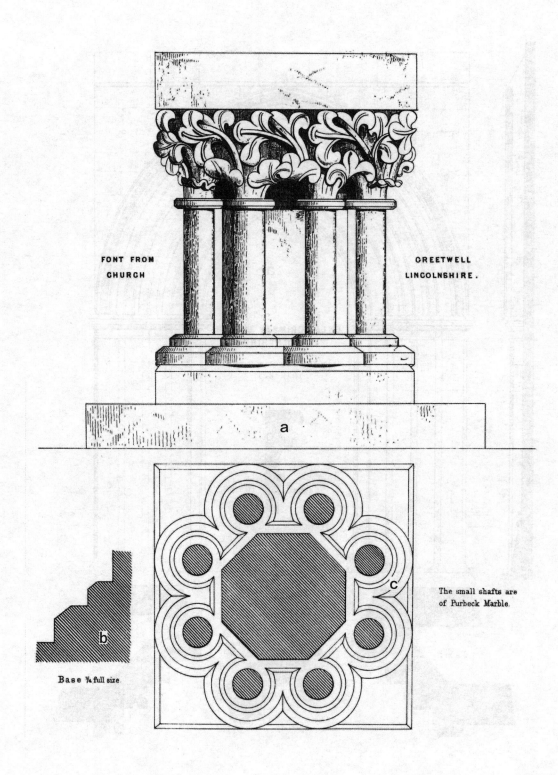

FONT FROM
CHURCH

GREETWELL
LINCOLNSHIRE.

a

b

Base ¾ full size.

c

The small shafts are
of Purbeck Marble.

林肯郡克里特维尔教堂：a. 洗礼池　b. 柱础　c. 波贝克大理石柱身

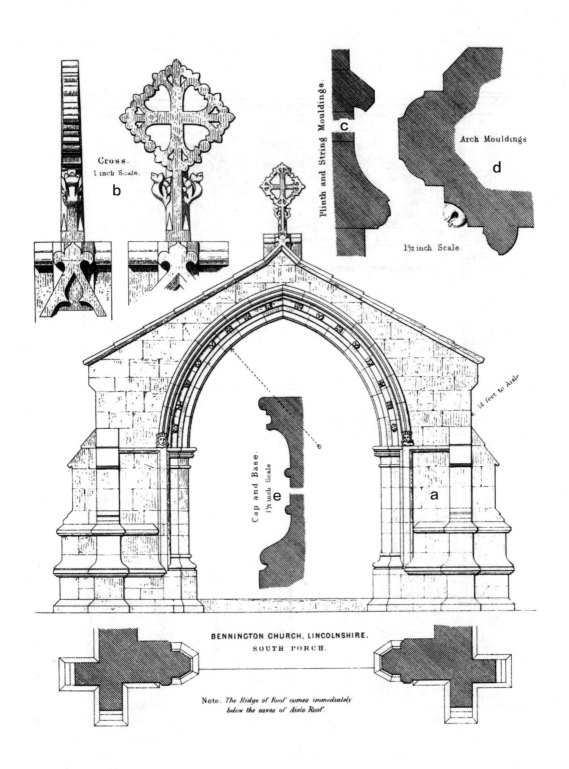

林肯郡本宁顿教堂：a. 南侧门廊　b. 十字架　c. 底座和束带层线脚　d. 拱线脚　e. 柱头和柱础　注意：屋脊线紧接在走廊屋脊下方

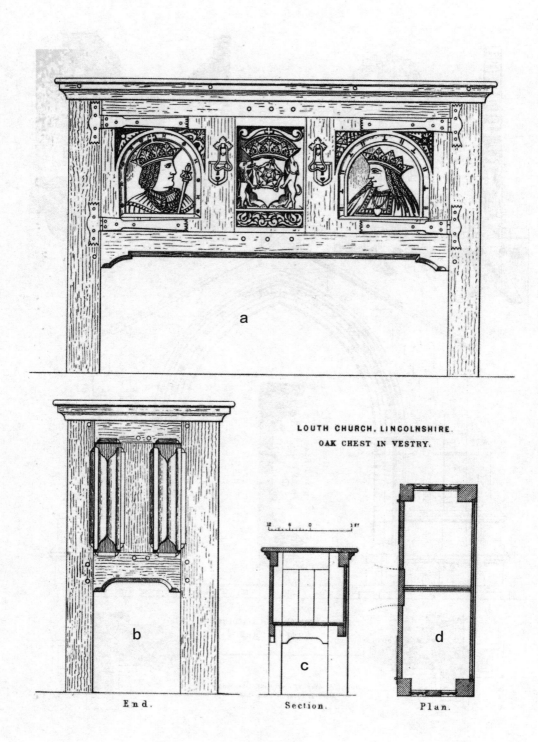

LOUTH CHURCH, LINCOLNSHIRE.
OAK CHEST IN VESTRY.

End.　　　Section.　　　Plan.

林肯郡劳斯教堂：a. 法衣室橡木的柜子　　b. 法衣室橡木柜子的侧面　　c. 法衣室橡木柜子的剖面　　d. 法衣室橡木柜子的平面

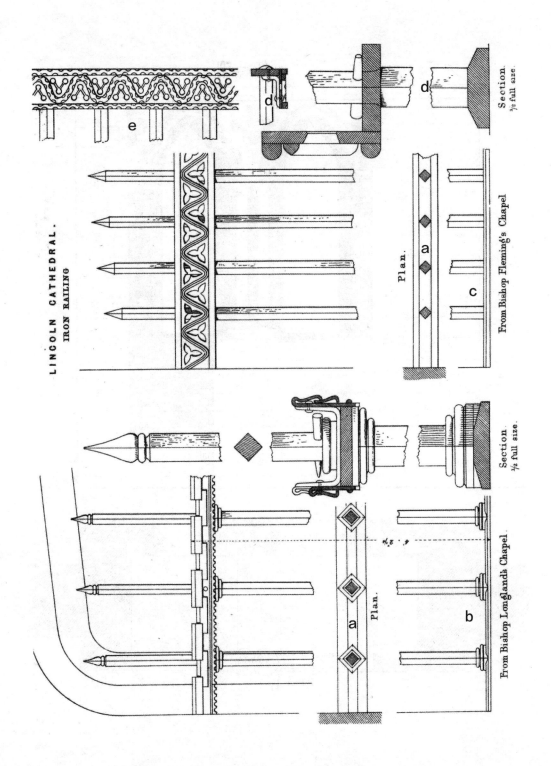

林肯大教堂铁栏杆：a. 平面　b. 位于郎兰主教堂里　c. 位于弗莱明主教堂里　d. 剖面　e. 拉塞尔主教堂下横档

林肯郡图弗尔密修道院：a. 餐室中的布道坛　b. 座位　c. 平面

TUPHOLME PRIORY

DETAILS OF PULPIT

LINCOLNSHIRE.

IN REFECTORY.

Sketch of
Centre Capitals.

a

Corbels.

d

b

Profile of
Capital and Base.

c

Arch Mouldings.

¼ full size.

林肯郡图弗尔密修道院餐室中布道坛细部：a. 中心柱子素描　　b. 柱头和柱础轮廓　　c. 拱线脚　　d. 梁托

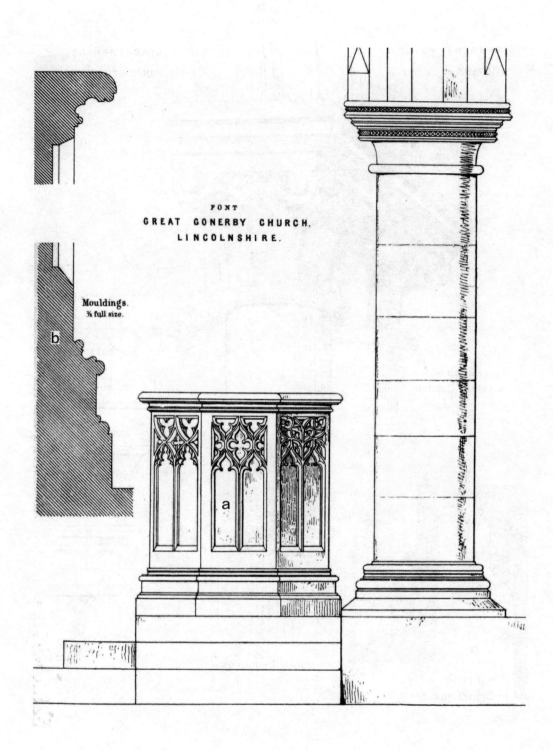

FONT
GREAT GONERBY CHURCH,
LINCOLNSHIRE.

Mouldings.
¾ full size.

林肯郡大戈纳比教堂：a. 洗礼盆　b. 线脚

GREAT GONERBY CHURCH, LINCOLNSHIRE.
PANELLING FROM FONT.

¼ full size.

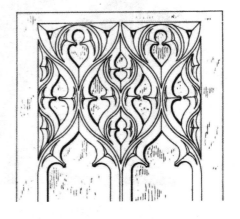

林肯郡大戈纳比教堂洗礼盆上的嵌板

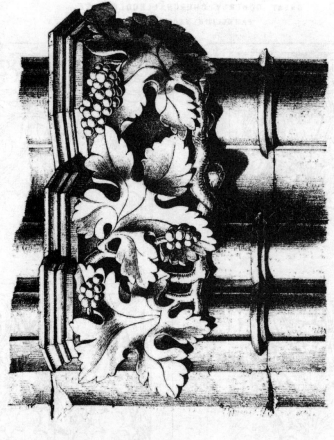

Capitals from the Cloisters, Lincoln Cathedral.
Half full size.

林肯大教堂回廊内柱头

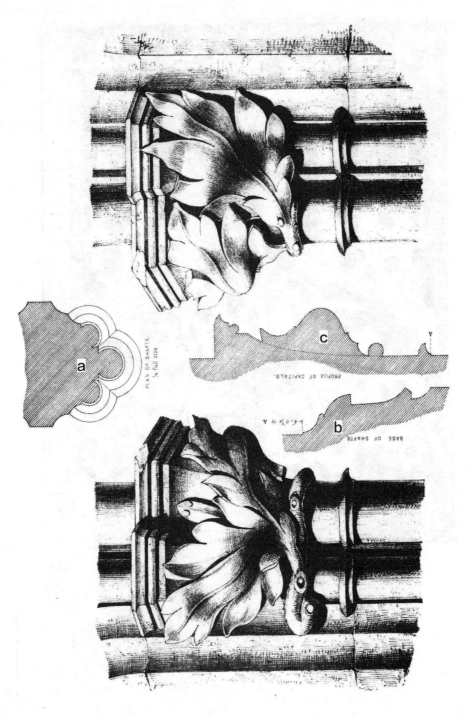

林肯大教堂回廊内柱头：a. 柱身平面　　b. 柱础　　c. 柱头轮廓

林肯郡塔特歇尔教堂：a. 布道坛上木质嵌板　　b. 布道坛 A 处的转角

林肯教堂唱诗室南走廊石制柱头：a. 柱顶板轮廓　　b. A 处的叶片装饰　　c. 平面

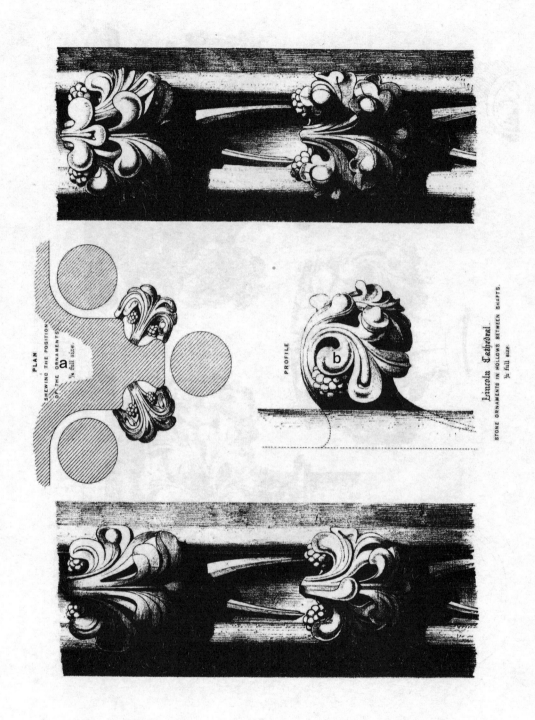

林肯教堂柱子间凹槽处石制装饰：a. 平面展示出装饰所处的位置　b. 轮廓

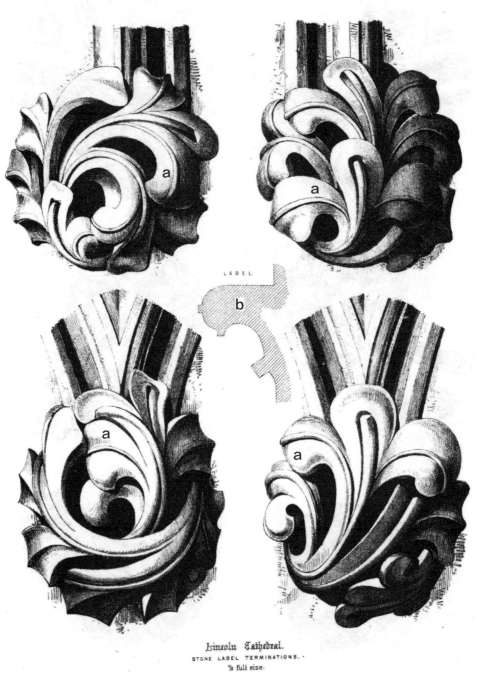

Lincoln Cathedral.
STONE LABEL TERMINATIONS.
½ full size.

林肯教堂：a. 披水石端部　　b. 披水石

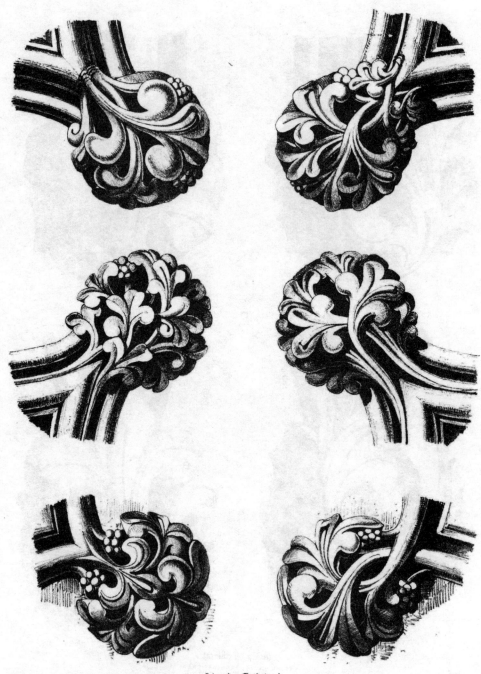

Lincoln Cathedral.

林肯教堂石制叶片状尖饰

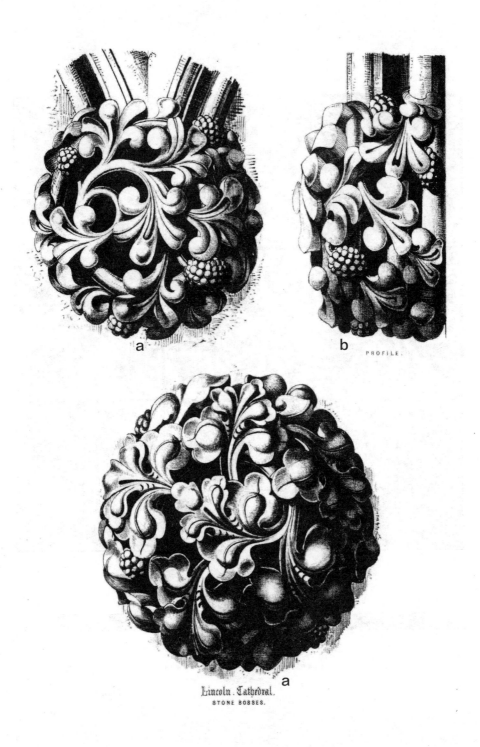

林肯教堂：a. 石制圆形浮雕　　b. 侧面

a

SECTION.

b

Lincoln Cathedral.

PORTION OF STONE DIAPER WORK.

½ full size.

林肯教堂：a. 石制菱形花饰局部　　b. 剖面

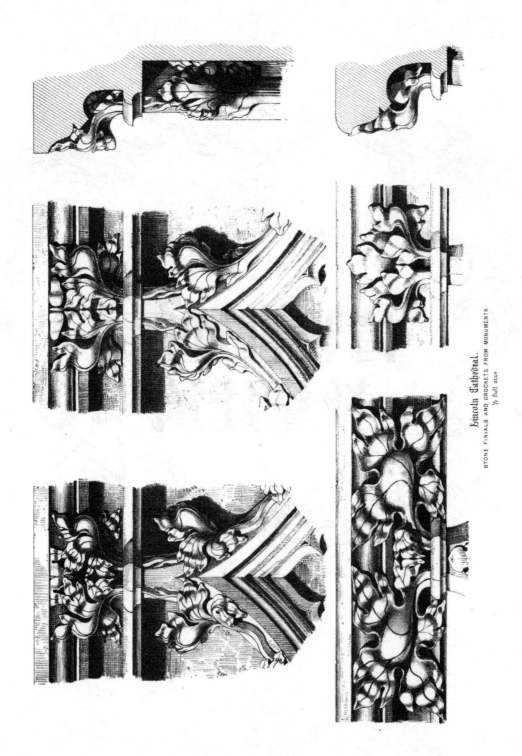

Lincoln Cathedral.

STONE FINIALS AND CROCKETS FROM MONUMENTS

½ full size.

林肯教堂遗迹上石制尖饰和卷叶形花饰

第 11 章

# 肯特郡地区教堂建筑

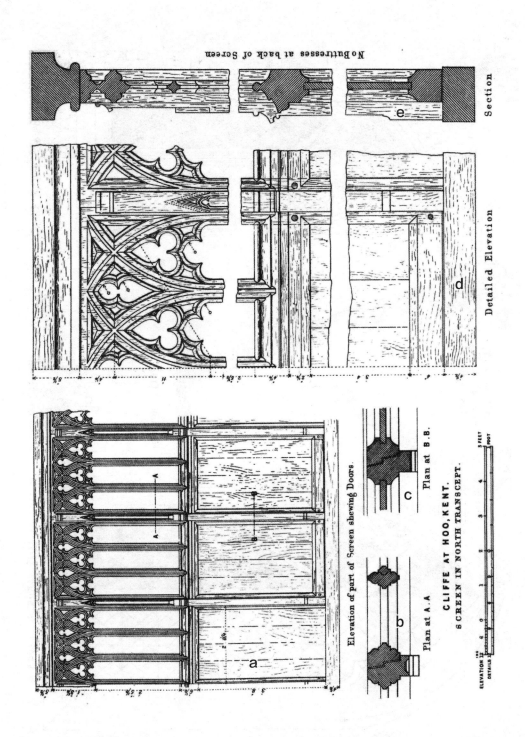

肯特郡克利夫教堂十字翼部隔断：a. 隔断上装饰性门局部立面　b.A处的平面　c.B处的平面　d. 立面细部　e. 剖面　隔断背面没有支撑

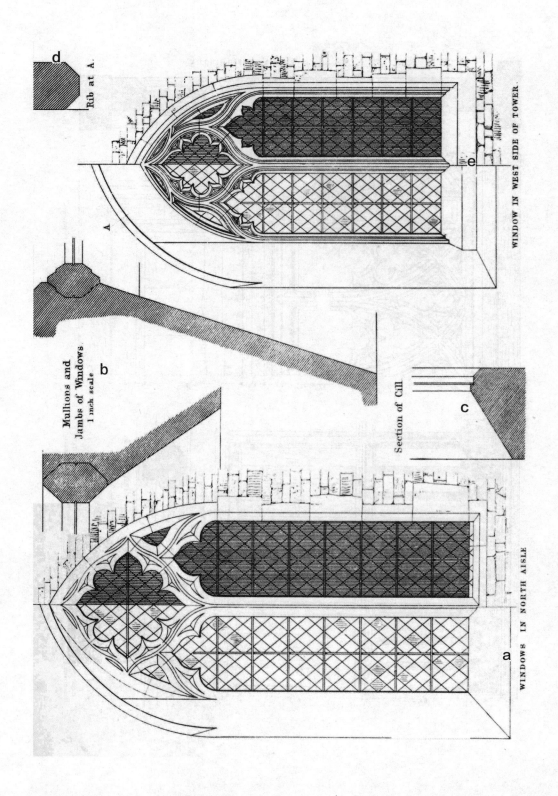

肯特郡罗尔文登教堂：a. 北侧走廊窗户　　b. 窗户竖框及侧壁　　c. 窗台剖面　　d.A 处的肋剖面　　e. 塔楼西侧窗户

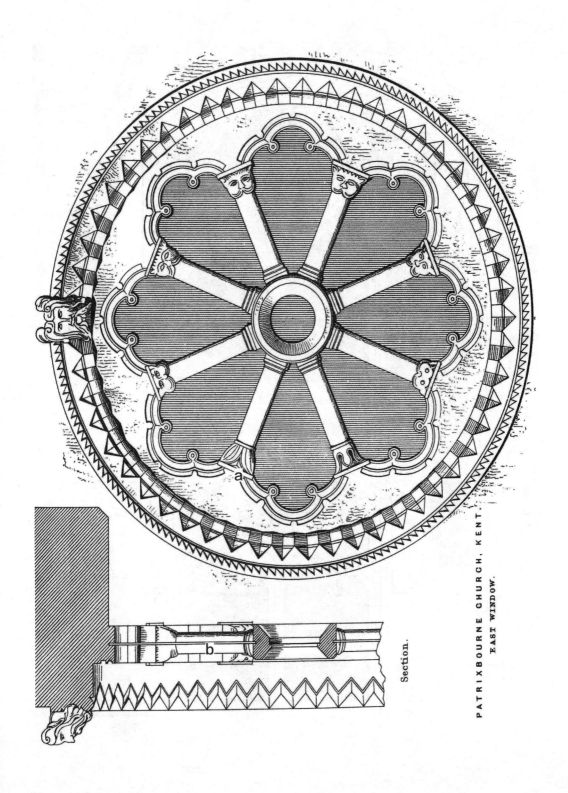

PATRIXBOURNE CHURCH, KENT.
EAST WINDOW.

Section.

肯特郡帕崔克死伯恩教堂：a. 东窗　b. 剖面

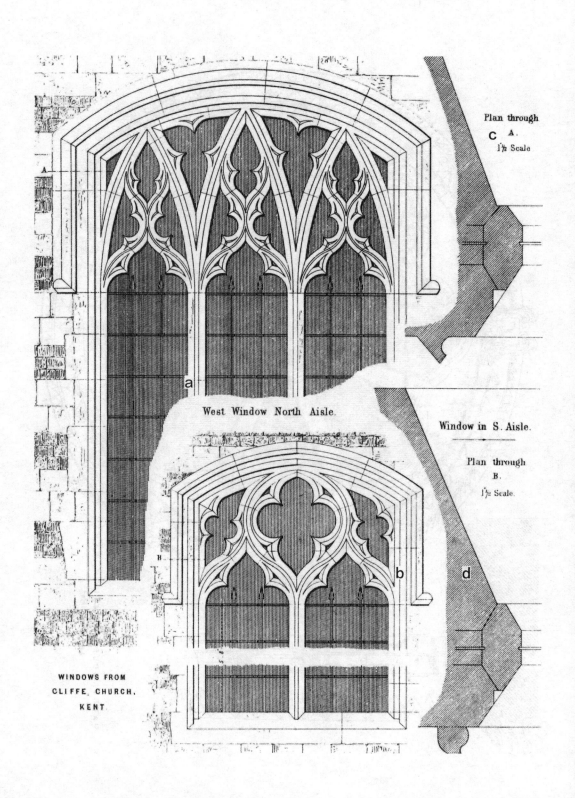

West Window North Aisle.

Plan through
C A.
1½ Scale

Window in S. Aisle.

Plan through
B.
1½ Scale

WINDOWS FROM
CLIFFE CHURCH.
KENT.

肯特郡克利夫教堂窗户：a. 北部走廊西侧窗户　b. 南部走廊窗户　c.A 处的平面　d.B 处的平面

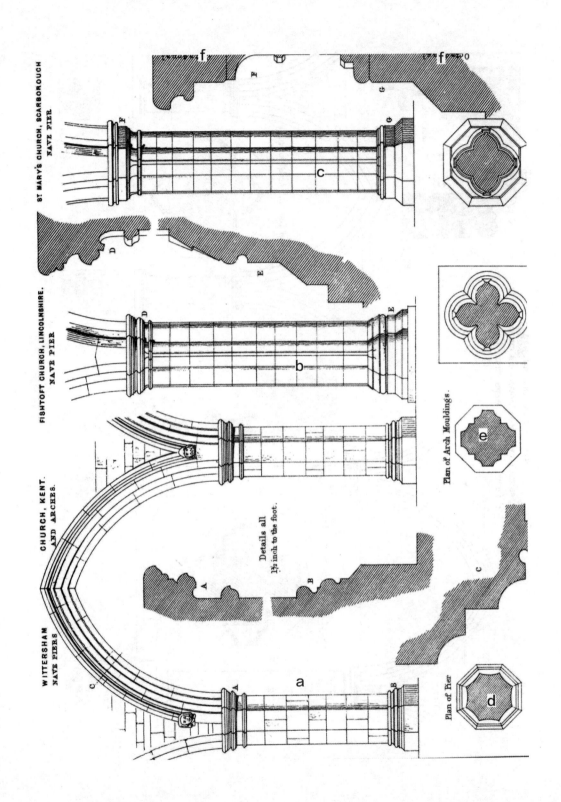

a. 肯特郡威特莎姆教堂中殿集束柱和拱　b. 林肯郡费西托夫特教堂中殿集束柱　c. 斯卡伯勒圣玛丽教堂中殿集束柱整体细部　d. 柱子平面　e. 拱线脚平面　f. 八边形

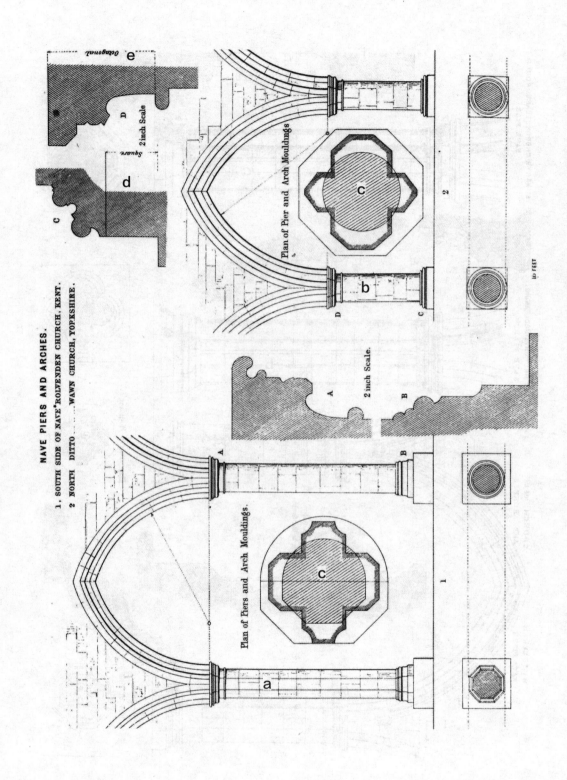

中殿集束柱和拱：a. 肯特郡罗尔文登教堂中殿南侧　b. 约克郡万恩教堂中殿北侧　c. 集束柱和拱平面　d. 方形　e. 八角形

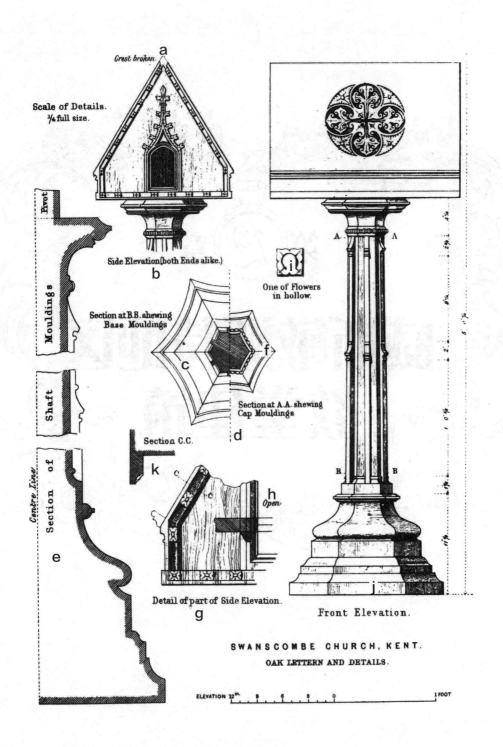

肯特郡斯旺斯孔教堂橡木诵经台及细部：a.损坏的顶饰　b.侧立面（两侧相似）　c.展示柱础线脚的B-B剖面　d.中心线　e.柱子剖面　f.展示柱头线脚的A-A剖面　g.侧立面局部细节　h.开启　i.凹处花饰　j.正立面　k.C-C剖面

# 第12章

# 威斯敏斯特地区教堂建筑

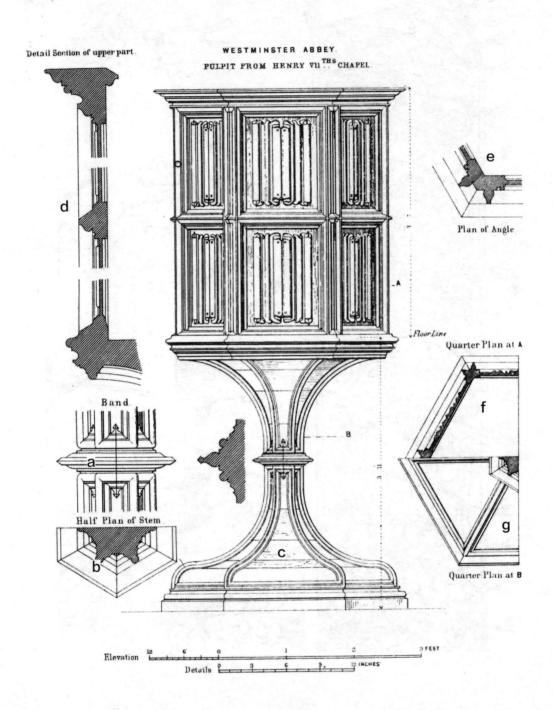

威斯敏斯特教堂亨利七世小礼拜堂讲道坛：a. 束环　b. 讲坛底柱平面一半　c. 立面　d. 上部详细剖面　e. 转角平面　f.A 处的四分之一平面　g.B 处的四分之一平面

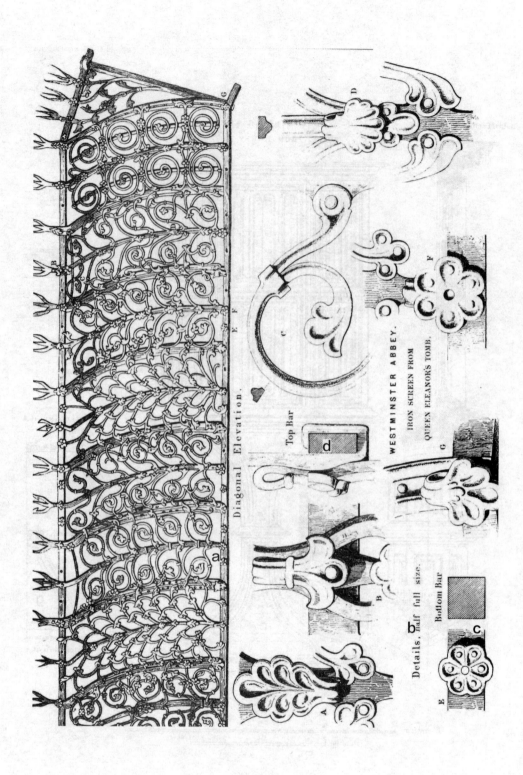

威斯敏斯特修道院埃莉诺女王墓碑上铁屏：a. 斜视立面　b. 细部　c. 底部铁条　d. 顶部铁条

威斯敏斯特修道院圣埃德蒙礼拜堂威廉德瓦伦西亚（彭布罗克郡伯爵，死于1296年）墓穴上镀金、铜菱形花饰样本：

a. 来自于墓穴表面　　b. 来自于尸体头部下面枕头上

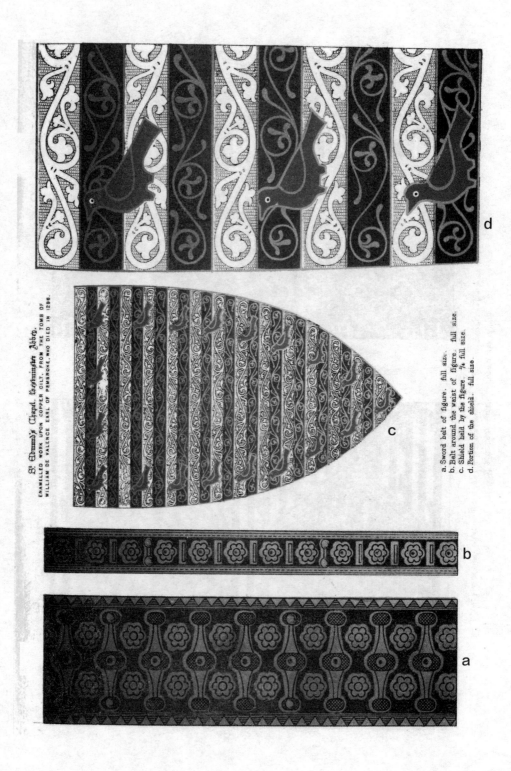

威斯敏斯特修道院圣埃德蒙礼拜堂威廉德瓦伦西亚（彭布罗克郡伯爵，死于 1296 年）墓穴上镀金、铜装饰：a. 佩剑腰带　b. 腰带　c. 盾牌　d. 盾牌局部

威斯敏斯特修道院回廊门口石制柱头注：柱颈、柱顶圆盘以及植物装饰上有红色残留的痕迹，表明植物装饰在过去可能有镀金

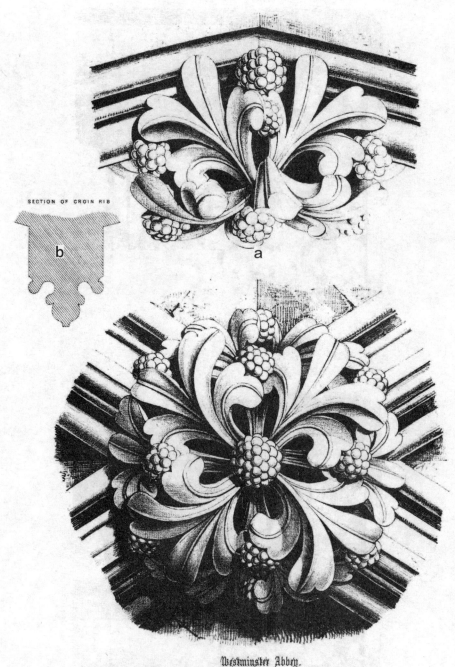

SECTION OF GROIN RIB

b

a

Westminster Abbey.
STONE BOSS FROM THE PASSAGE LEADING TO THE CHAPTER HOUSE.
⅓rd full size.

威斯敏斯特修道院：a. 通向牧师会礼堂走廊上的石制凸饰　b. 穹棱剖面

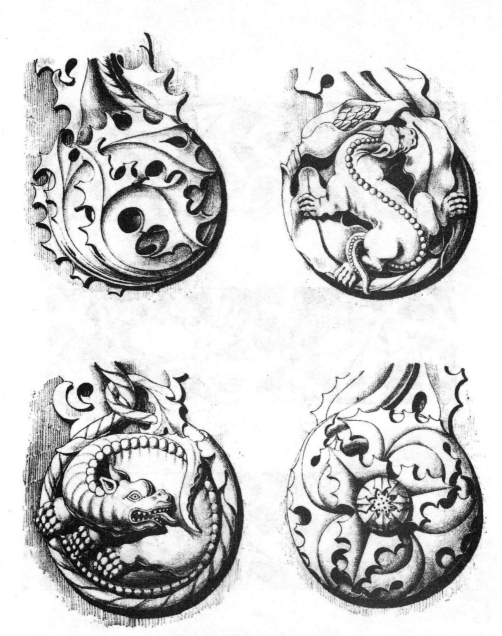

Henry VII.ᵗʰ Chapel, Westminster Abbey.
SUBSELLÆ FROM THE STALLS.
½ full size.

威斯敏斯特修道院亨利七世圣堂座椅上的圆形凸饰

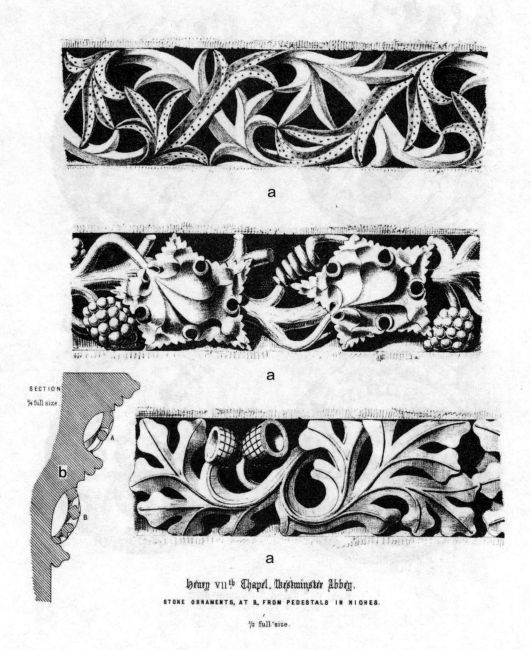

Henry VIIᵗʰ Chapel. Westminster Abbey.

STONE ORNAMENTS, AT B, FROM PEDESTALS IN NICHES.

½ full size.

威斯敏斯特修道院亨利七世圣堂：a. 壁龛基座上的石制装饰　b. 剖面

Henry vii<sup>th</sup> Chapel. Westminster Abbey.

STONE ORNAMENTS. AT A, PLATE 52, FROM PEDESTALS IN NICHES.

½ full size.

威斯敏斯特修道院亨利七世圣堂壁龛基座上的石制装饰

威斯敏斯特修道院圣伊拉兹马斯圣殿内石制拱肩

束带层上石制花饰：a、b、c、d. 来自威斯敏斯特修道院阿伯特·福西特纪念碑　　e、f. 来自萨福克拉文姆教堂

第 13 章

# 温彻斯特地区教堂建筑

a

b

ORNAMENTAL TILES.

¼ full size.

c

a、b、d、e. From St Cross. near Winchester.

c. Tamworth Church, Staffordshire.

f. Winchester Cathedral.

d

e

f

装饰砖：a、b、d、e. 来源于温彻斯特附近的圣十字教堂　c. 来自斯塔福德郡塔姆沃斯教堂　f. 来自温彻斯特大教堂

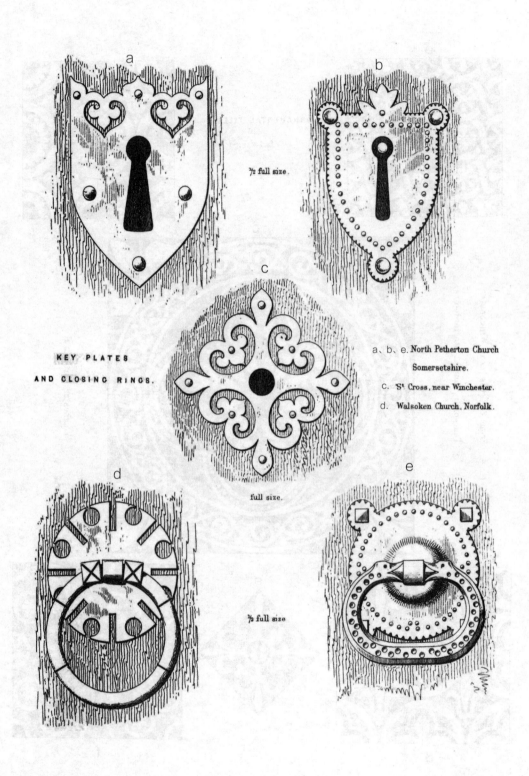

KEY PLATES
AND CLOSING RINGS.

a、b、e. North Petherton Church
Somersetshire.

c. S⁺ Cross, near Winchester.

d. Walsoken Church, Norfolk.

½ full size.

full size.

½ full size

钥匙孔板和拉环：a、b、e.来自索美赛特夏郡佩瑟顿教堂　c.来自温切斯特附近圣十字教堂　d.来自诺福克郡沃尔索肯教堂

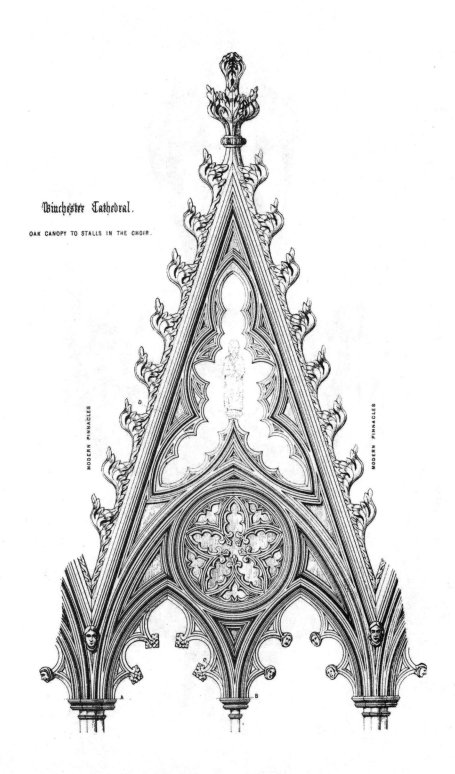

温彻斯特教堂唱诗室内座椅橡木顶棚

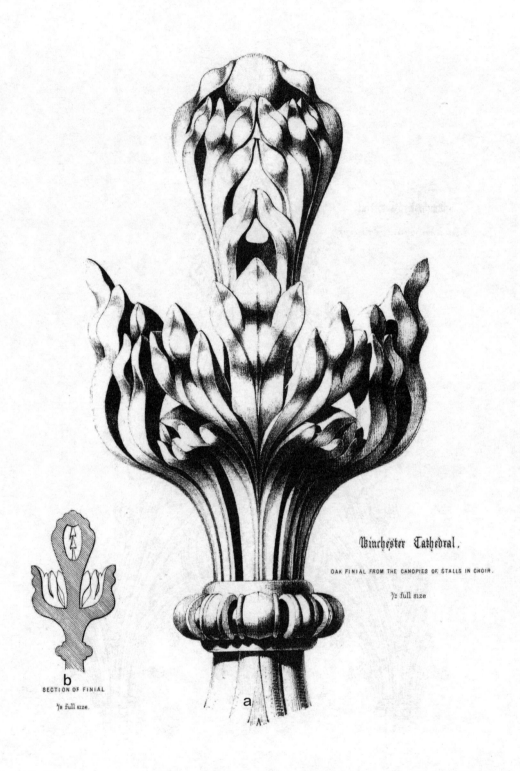

温彻斯特教堂：a. 唱诗室内座椅橡木顶棚尖顶饰　b. 尖顶饰剖面

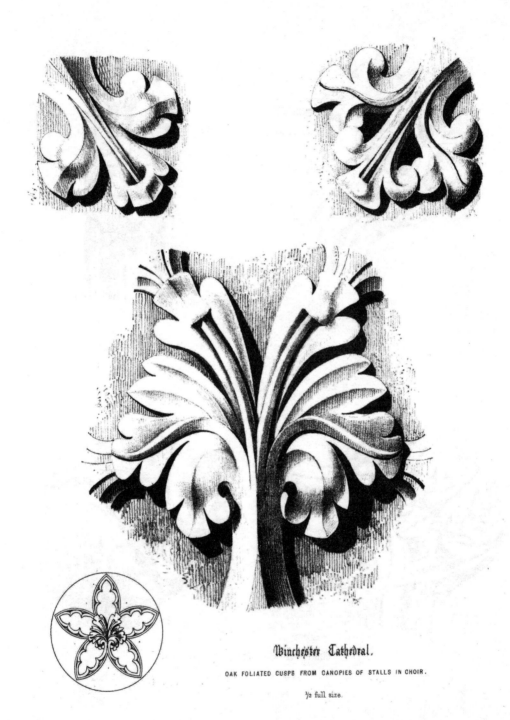

Winchester Cathedral.

OAK FOLIATED CUSPS FROM CANOPIES OF STALLS IN CHOIR.

½ full size.

温彻斯特教堂唱诗室内座椅橡木顶棚叶形饰尖头

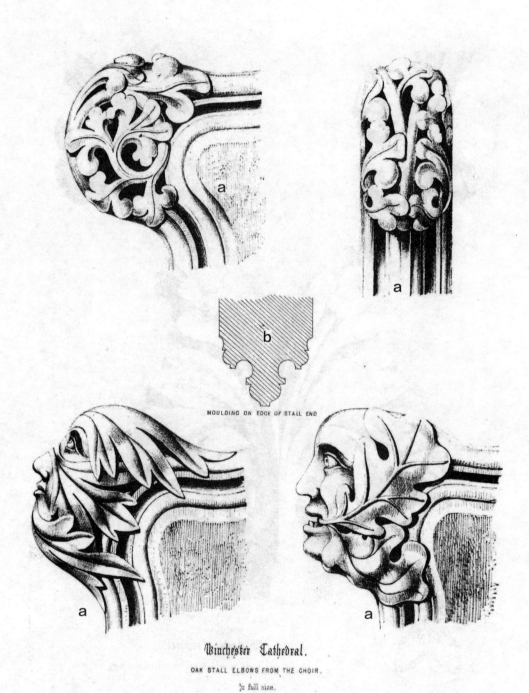

Winchester Cathedral.

OAK STALL ELBOWS FROM THE CHOIR.

½ full size.

温彻斯特教堂：a. 唱诗室内座椅的橡木扶手　　b. 座椅的端部边缘线脚

温彻斯特教堂唱诗室内座椅上的橡木隔断上的拱形装饰

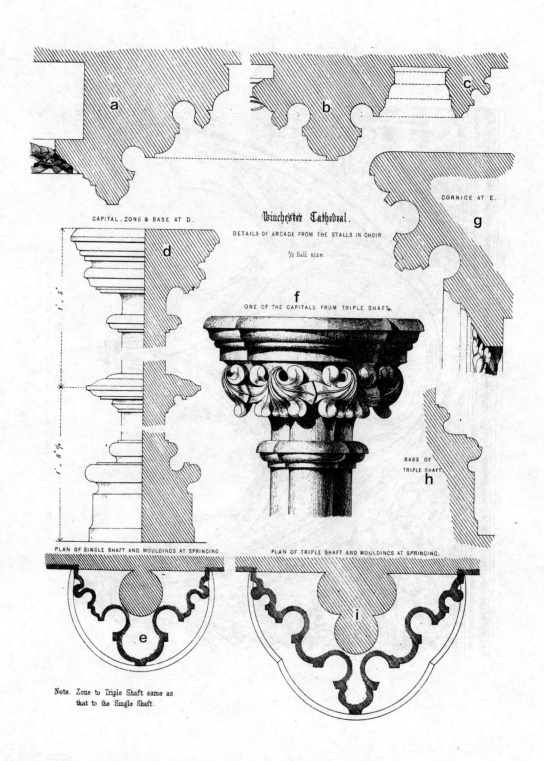

温彻斯特教堂唱诗室内座椅上橡木隔断上拱装饰细部：a.A 处的线脚　　b.B 处的线脚　　c.C 处的线脚　　d.D 处的柱头、束带及柱础　　e.单柱平面及束带处线脚　　f.三颗柱子之一的柱头　　g.E 处的檐口　　h.三柱柱础　　i.三柱平面及束带处线脚

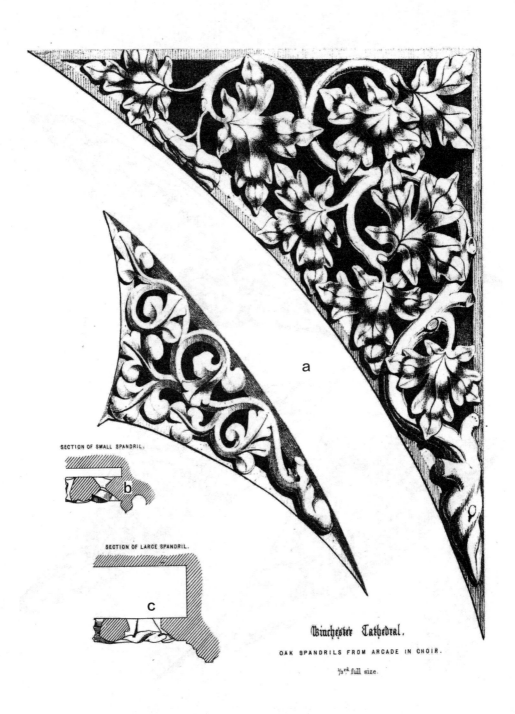

SECTION OF SMALL SPANDRIL.

SECTION OF LARGE SPANDRIL.

Winchester Cathedral.

OAK SPANDRILS FROM ARCADE IN CHOIR.

⅓rd full size.

温彻斯特教堂：a. 唱诗室内的橡木拱肩　b. 小拱肩剖面　c. 大拱肩剖面

温彻斯特教堂：a. 唱诗室内的橡木拱肩　　b. 剖面

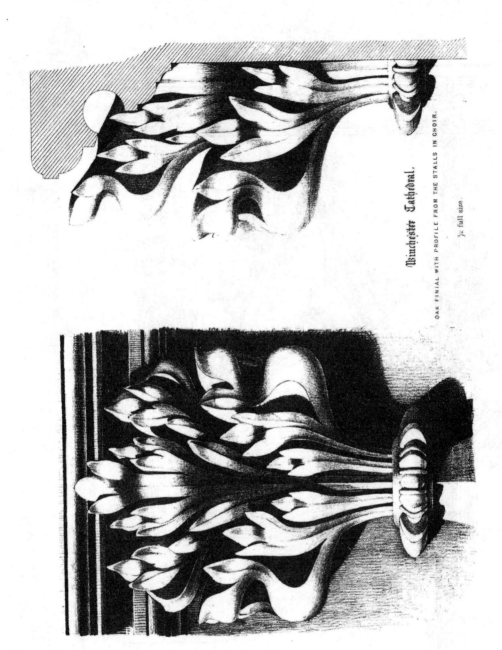

温彻斯特教堂唱诗室内座椅尖顶饰

温彻斯特教堂长老会内拱廊上石制顶棚：a. 剖面　　b.C 处的平面　　c.B 处的线脚剖面

Winchester Cathedral.

STONE FINIAL, AT A PLATE 92.

½ full size.

温彻斯特教堂石制尖顶饰

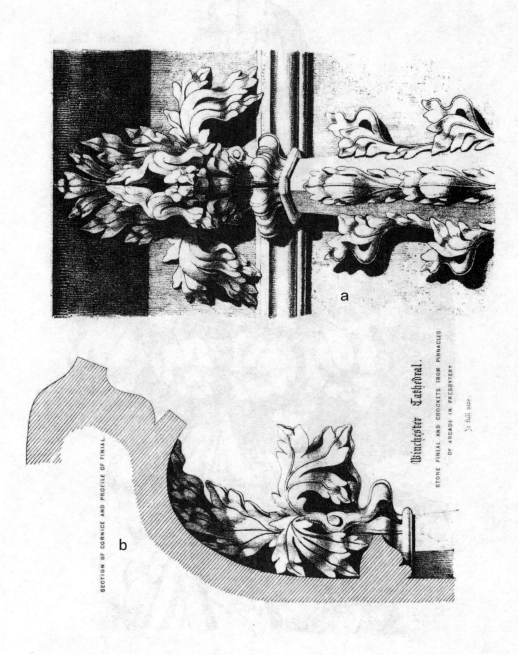

温彻斯特教堂：a. 长老会内拱廊上小尖塔石制尖顶饰及卷叶形花饰　　b. 檐口剖面及尖顶饰侧面

温彻斯特教堂：a. 长老会拱廊石制凸饰　　b. 肋剖面　　c. 侧面

温彻斯特教堂：a. 长老会拱廊线脚凹槽处石制装饰    b. 剖面

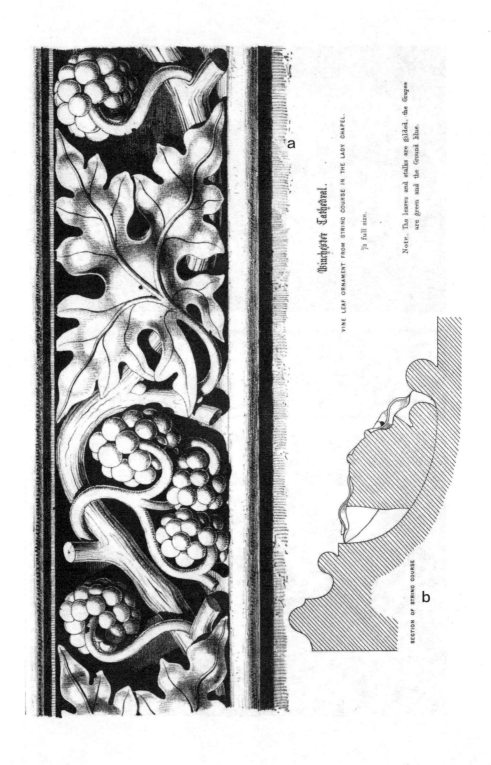

温彻斯特教堂：a. 圣母堂内束带层上葡萄叶饰　　b. 束带层剖面

温彻斯特教堂长老会内铺地的瓷砖样本

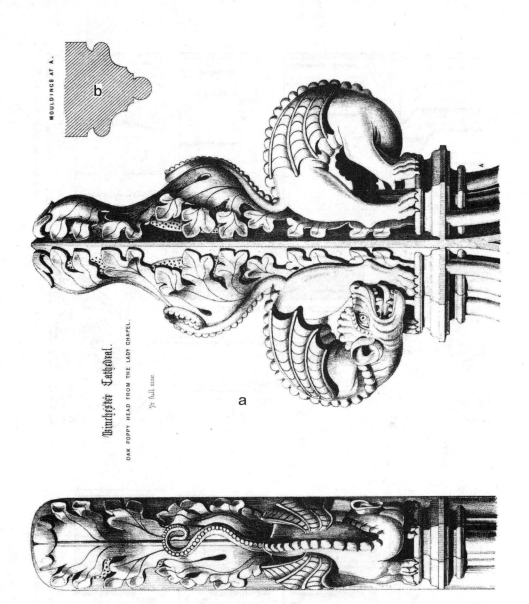

Winchester Cathedral.

OAK POPPY HEAD FROM THE LADY CHAPEL.

½ full size.

MOULDINGS AT A.

温彻斯特教堂：a. 圣母堂内橡木罂粟状装饰　　b.A 处的线脚

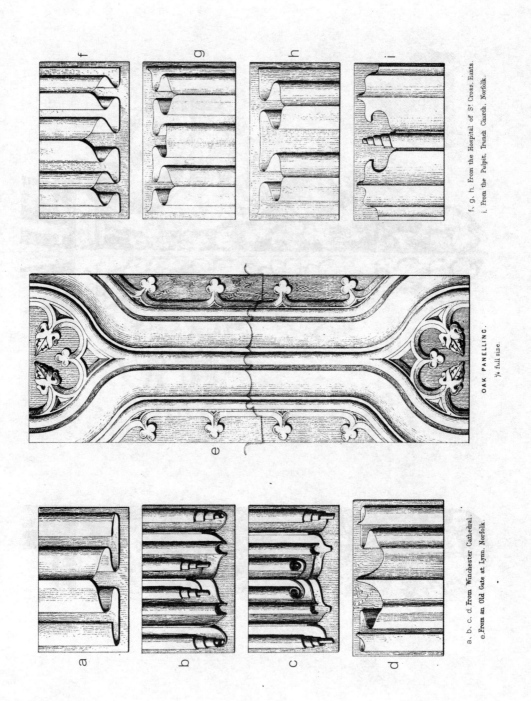

橡木嵌板：a、b、c、d. 来自温彻斯特教堂　e. 来自诺福克郡林恩市一个古门　f、g、h. 来自汉普郡圣十字医院 i. 来自创奇教堂布道台

# 第14章

# 以利地区教堂建筑

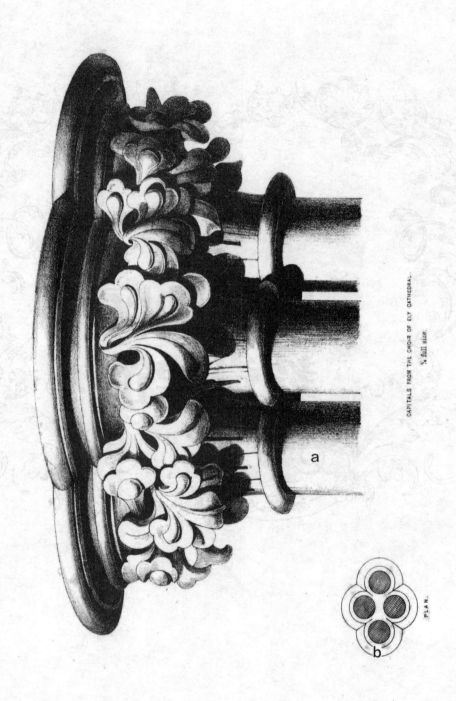

CAPITALS FROM THE CHOIR OF ELY CATHEDRAL.

¾ full size.

a

PLAN.

b

以利教堂：a. 唱诗室柱头　　b. 平面

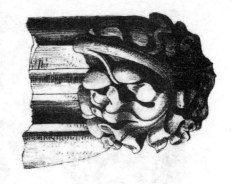

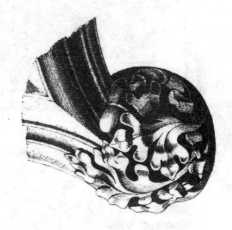

以利教堂牧师会礼堂里石制叶片尖端装饰

PROFILE
⅓ full size.

Ely Cathedral
KNOTS OF FOLIAGE FROM THE CHAPTER HOUSE

以利教堂：a. 牧师会礼堂内结状叶子装饰    b. 侧面

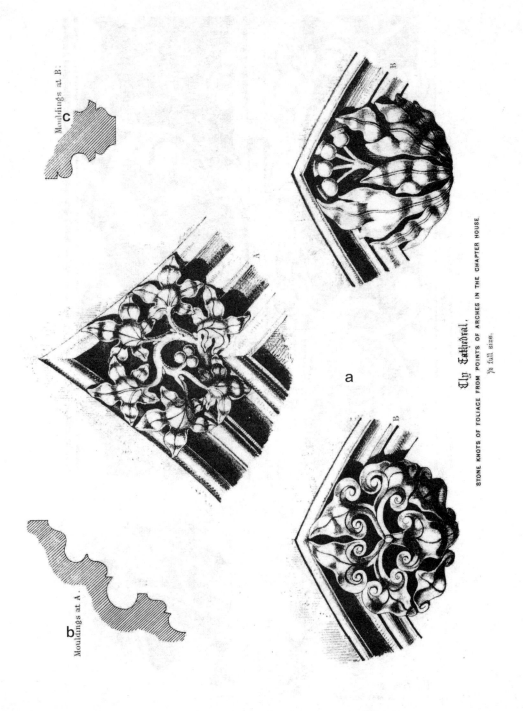

以利教堂：a. 牧师会礼堂内拱顶结状叶子装饰　　b.A 处的线脚　　c.B 处的线脚

Ely Cathedral.
ORNAMENT FROM ARCH OF WEST DOORWAY.

以利教堂西门拱上的装饰

以利教堂：a. 圣母堂内座椅石制顶棚　　b. 座位平面

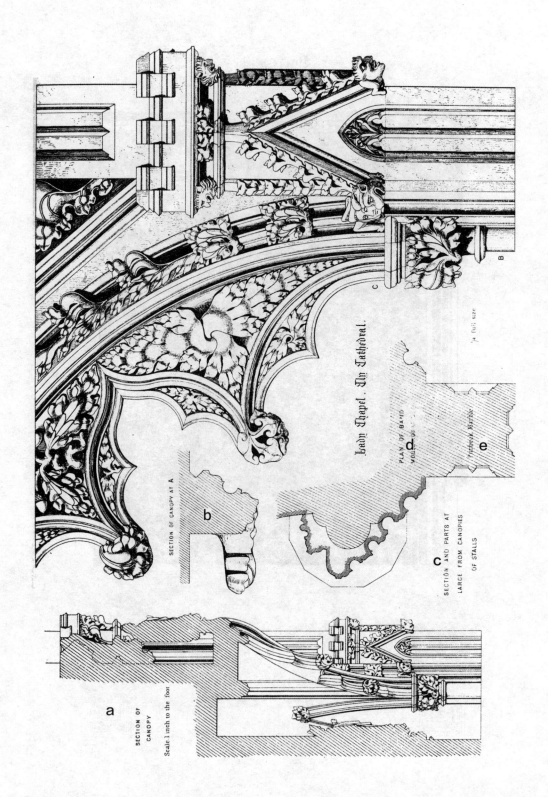

以利教堂圣母堂：a. 顶棚剖面　　b.A 处的顶棚剖面　　c. 座椅顶棚剖面及部件详细图　　d.C 处的圆盘线脚平面
e. 波贝克大理石

Lady Chapel. Ely Cathedral.

STONE CAPITALS FROM STALLS

以利教堂圣母堂座椅上的石制柱头

以利普赖尔克兰登教堂装饰性瓷砖铺面局部（瓷砖上蔷薇纹饰是雕刻的）

以利普赖尔克兰登教堂：a. 装饰性瓷砖铺面局部　　b. 组成铺面的瓷砖表面上的线条是雕刻的

第15章

# 诺丁汉郡地区
# 教堂建筑

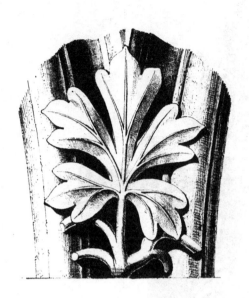

*Southwell Minster, Notts.*

STONE ORNAMENTS AT THE SPRINCING OF ARCHES OF ARCADE IN THE CHAPTER HOUSE.

½ full size.

诺丁汉郡索斯韦尔大教堂牧师会礼堂拱廊内起拱线处石制装饰

诺丁汉郡索斯韦尔大教堂牧师会礼堂拱廊顶棚局部：a.A 处的线脚　b.B 处的线脚

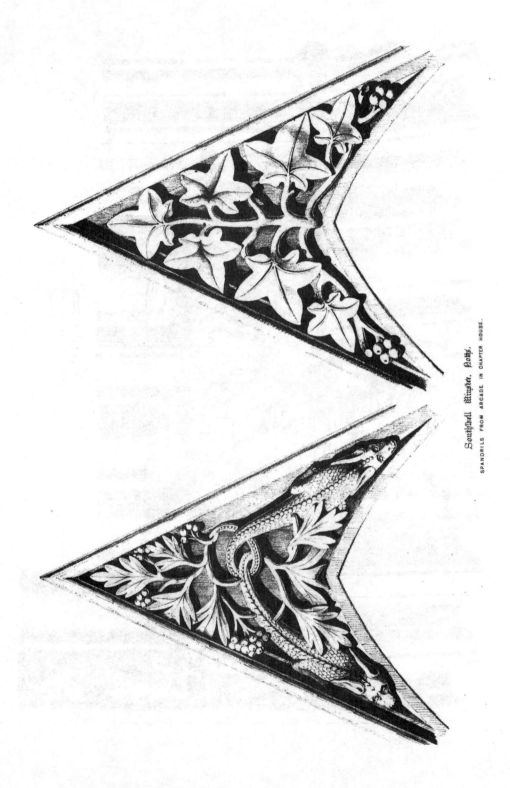

诺丁汉郡索斯韦尔大教堂牧师会礼堂拱廊内拱肩

诺丁汉郡索斯韦尔大教堂牧师会礼堂大门上石制柱头

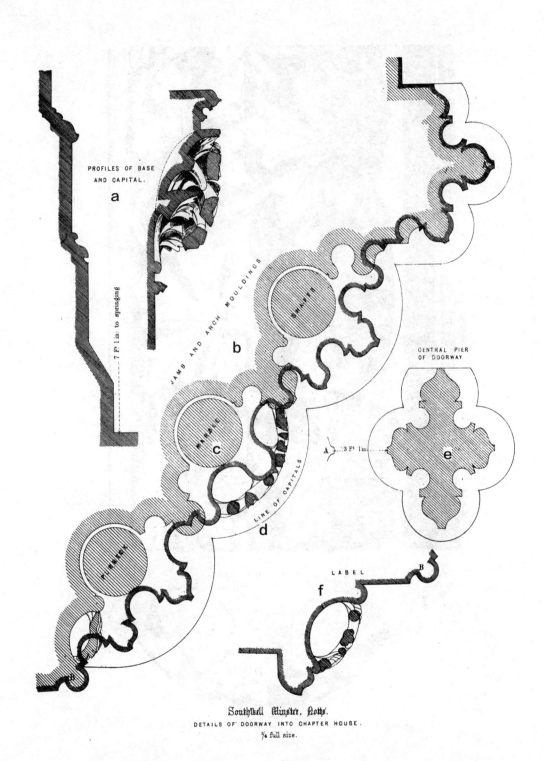

诺丁汉郡索斯韦尔大教堂牧师会礼堂大门细部：a. 柱础及柱头轮廓　　b. 侧壁及拱线脚　　c. 波贝克大理石柱身
d. 柱头投影线　　e. 大门中心柱子　　f. 披水石

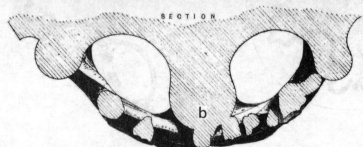

诺丁汉郡索斯韦尔大教堂：a. 牧师会礼堂大门拱上石制枫叶装饰　b. 剖面

Southwell Minster. Notts.

STONE VINE LEAF ORNAMENT FROM THE LABEL OF DOORWAY TO CHAPTER HOUSE.

½ full size.

诺丁汉郡索斯韦尔大教堂牧师会礼堂大门披水石上石制葡萄叶装饰

Southwell Minster, Notts.
STONE SPANDRILS FROM ARCADE IN CHAPTER HOUSE.
⅓ʳᵈ full size.

诺丁汉郡索斯韦尔大教堂牧师会礼堂拱廊上石制拱肩

Southwell Minster. Notts.

STONE BOSSES TERMINATING THE CANOPIES OF THE ARCADE IN THE CHAPTER HOUSE.

诺丁汉郡索斯韦尔大教堂牧师会礼堂拱廊顶部端头石制圆形浮雕

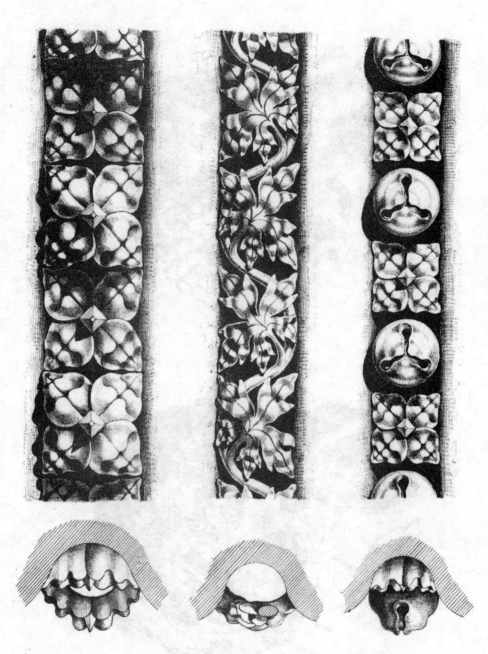

S! Albans' Abbey Church, Herts.

ORNAMENTS BETWEEN SHAFTS FROM THE TRIFORIUM OF NAVE.

⅓rd full size.

诺丁汉郡圣奥尔本斯修道院教堂中殿拱门上拱廊柱间装饰

诺丁汉郡圣奥尔本斯修道院教堂：a. 西面南侧门内石制托架　b. 柱顶板剖面

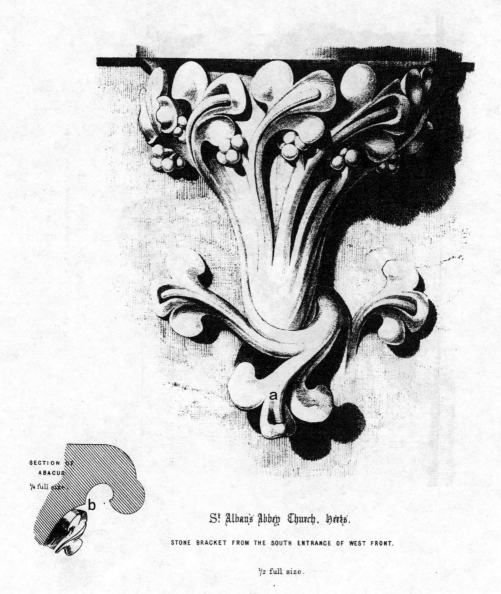

St Alban's Abbey Church. Herts.

STONE BRACKET FROM THE SOUTH ENTRANCE OF WEST FRONT.

½ full size.

诺丁汉郡圣奥尔本斯修道院教堂：a. 西面南侧入口石制托架　b. 柱顶板剖面

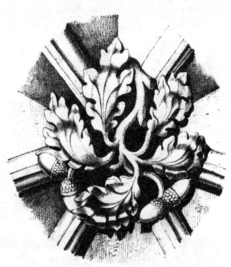

St. Alban's Abbey Church, Herts.

STONE BOSSES FROM ST CUTHBERT'S SCREEN.

¼ full·size

诺丁汉郡圣奥尔本斯修道院教堂圣库斯伯特屏风上的石制凸饰

诺丁汉郡圣奥尔本斯修道院教堂：a. 圣坛交叉穹顶木制天花板上的彩绘　b.A 处的花图样　c. 穹棱剖面　d.B 处的花图样

St. Alban's Abbey Church. Herts.

STONE CORBEL FROM THE SOUTH ENTRANCE OF WEST FRONT.

½ full size.

诺丁汉郡圣奥尔本斯修道院教堂西面南侧入口的石制牛腿

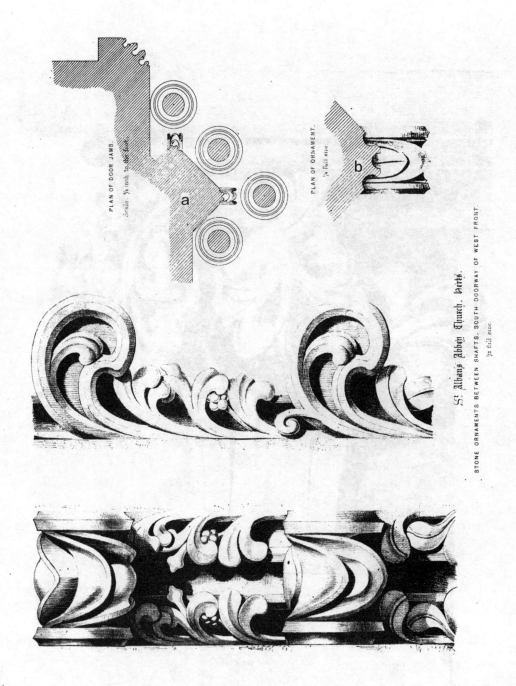

诺丁汉郡圣奥尔本斯修道院教堂西面南侧入口柱间石制装饰：a. 门侧壁平面　b. 装饰平面

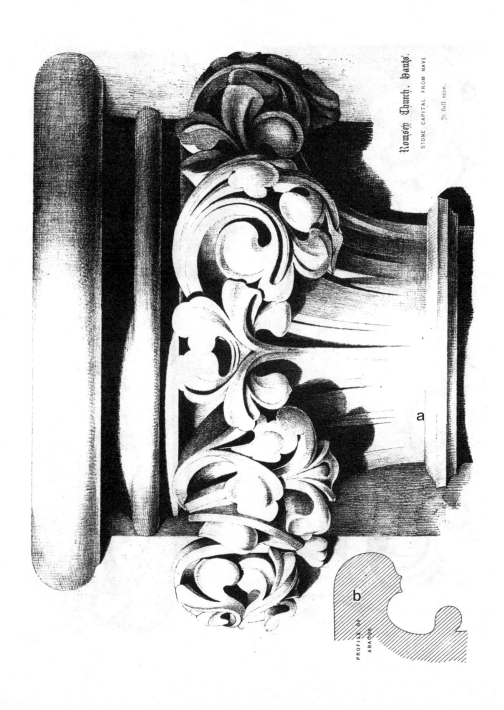

诺丁汉郡拉姆齐教堂：a. 中殿内石制柱头　b. 柱顶板轮廓

诺丁汉郡拉姆齐教堂：a. 中殿内束柱柱础上的石制装饰　b. 平面

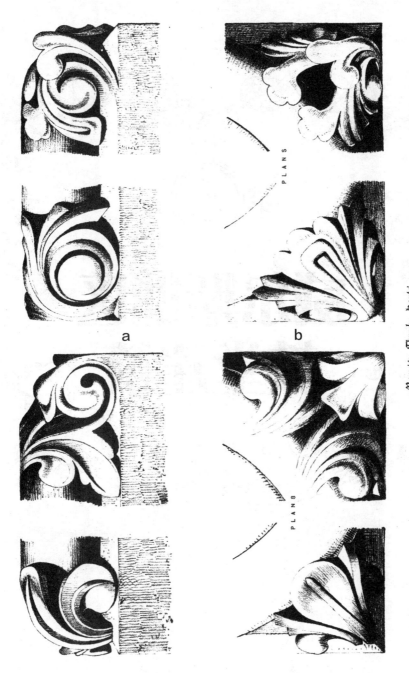

Rempson Church, Hants.

PLANS

PLANS

STONE ORNAMENTS TO ANGLES OF BASES OF PIERS FROM THE NAVE.

a

b

诺丁汉郡拉姆齐教堂：a. 中殿内束柱柱础转角处的石制装饰    b. 平面

第 16 章

# 威尔斯地区教堂建筑

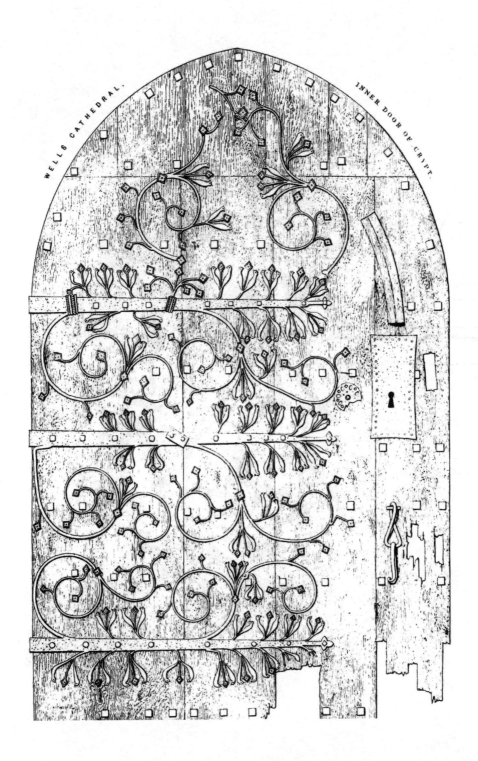

威尔斯大教堂地下室内门

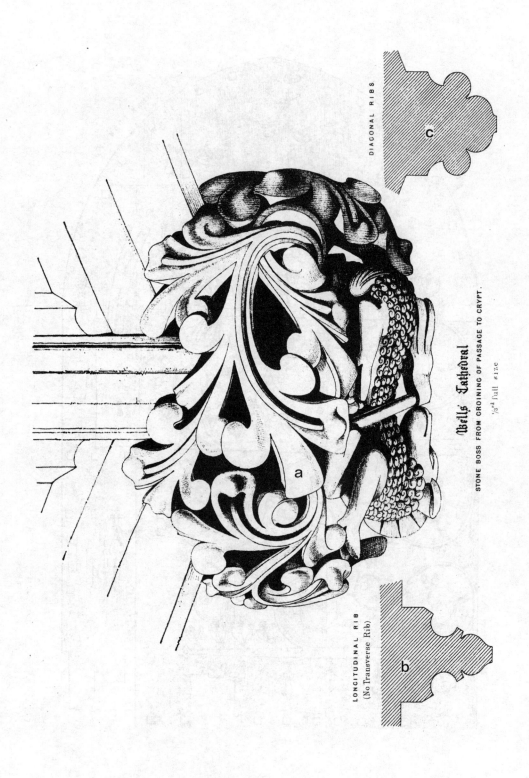

威尔斯教堂：a. 通向地下室通道穹顶上的石制凸饰　b. 纵向穹棱　c. 斜向穹棱

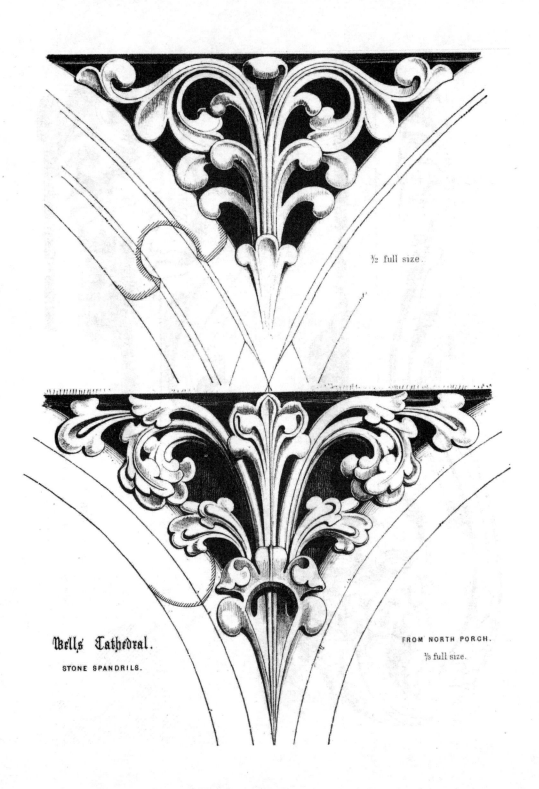

½ full size.

Wells Cathedral.

STONE SPANDRILS.

FROM NORTH PORCH.

⅓ full size.

威尔斯教堂北侧走廊内的石制拱肩

威尔斯教堂北侧走廊束带层端部

ORNAMENTAL TILES.
⅓ full size.

a、b、c、d、e. from Wells Cathedral
Grey & Yellow.
f、g. from Winchester Cathedral
Red & Yellow.

装饰瓷砖：a、b、c、d、e. 来自威尔斯教堂 灰色和黄色　f、g. 来自温切斯特教堂 红色和黄色

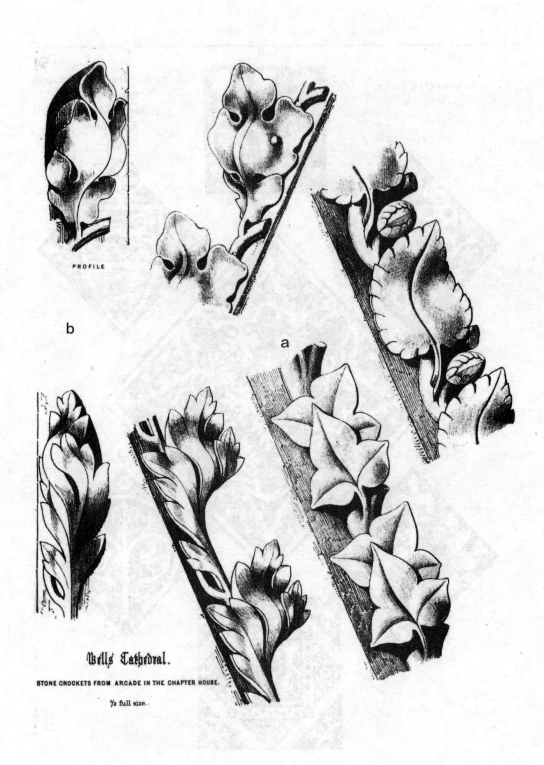

威尔斯教堂：a. 牧师会礼堂拱廊内石制卷叶形花饰　　b. 侧面

SECTION OF ABACUS

b

a

Wells Cathedral.

STONE CORBEL FROM PASSAGE TO CRYPT UNDER CHAPTER HOUSE

⅓ʳᵈ full size.

威尔斯教堂：a. 通向牧师会礼堂地下室通道内的石制托架　b. 柱顶板剖面

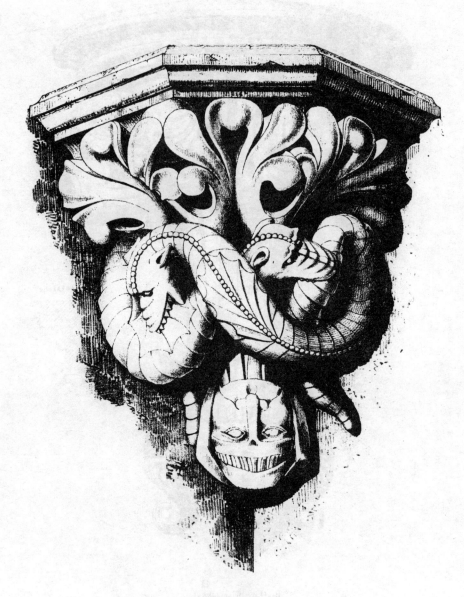

Wells Cathedral.

STONE ANGLE CORBEL FROM PASSAGE TO CRYPT UNDER CHAPTER HOUSE.

½ full size.

威尔斯教堂通向牧师会礼堂地下室通道内的带角托架

第 17 章

# 剑桥郡地区
# 教堂建筑

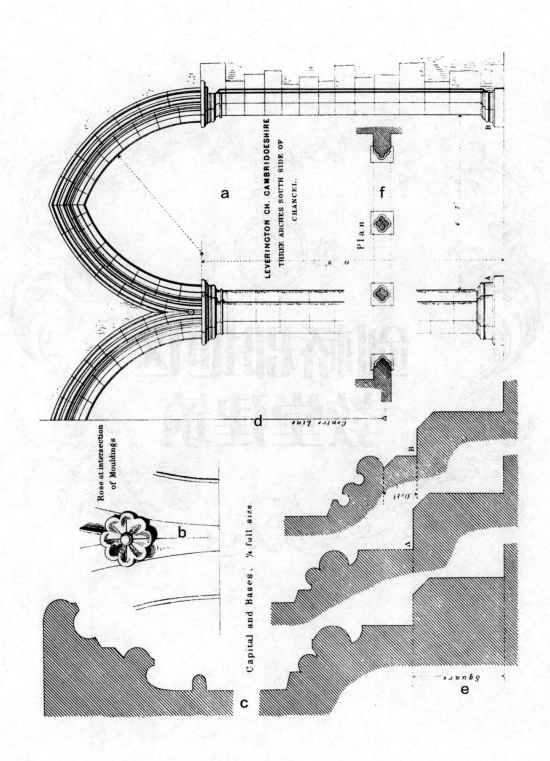

剑桥郡莱弗灵顿教堂：a. 圣坛南侧的三个拱门　b. 线脚交汇处玫瑰装饰　c. 柱头和柱础　d. 中心线　e. 正方　f. 平面

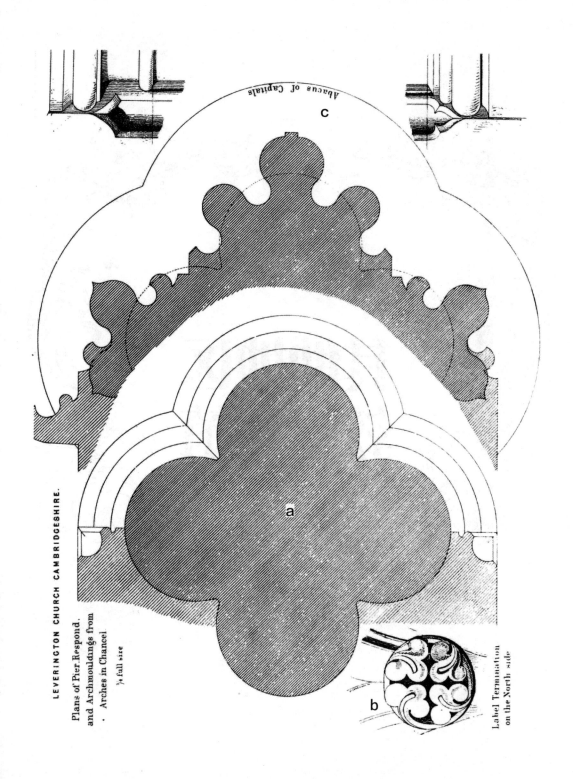

剑桥郡莱弗灵顿教堂：a. 圣坛拱门集束柱、壁联和线脚平面　b. 北侧披水石端部　c. 圆柱顶板

第 18 章

# 其他地区教堂建筑

SWINESHEAD CHURCH, HUNTINGDONSHIRE.
WEST OR TOWER ENTRANCE

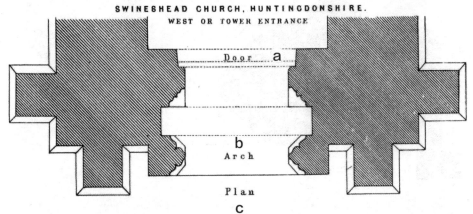

亨廷顿郡斯温斯黑德教堂西入口或塔楼入口：a. 门　b. 拱　c. 平面

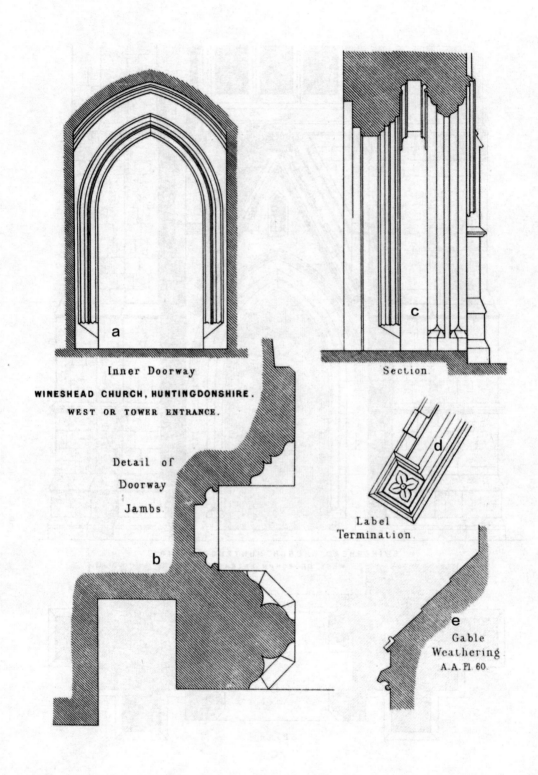

Inner Doorway

WINESHEAD CHURCH, HUNTINGDONSHIRE.

WEST OR TOWER ENTRANCE.

Detail of
Doorway
Jambs.

Section.

Label
Termination.

Gable
Weathering.
A.A. Pl. 60.

亨廷顿郡斯温斯黑德教堂西入口或塔楼入口：a. 大门内侧　b. 大门侧壁细部　c. 剖面　d. 披水石端部　e. 山墙泄水斜度

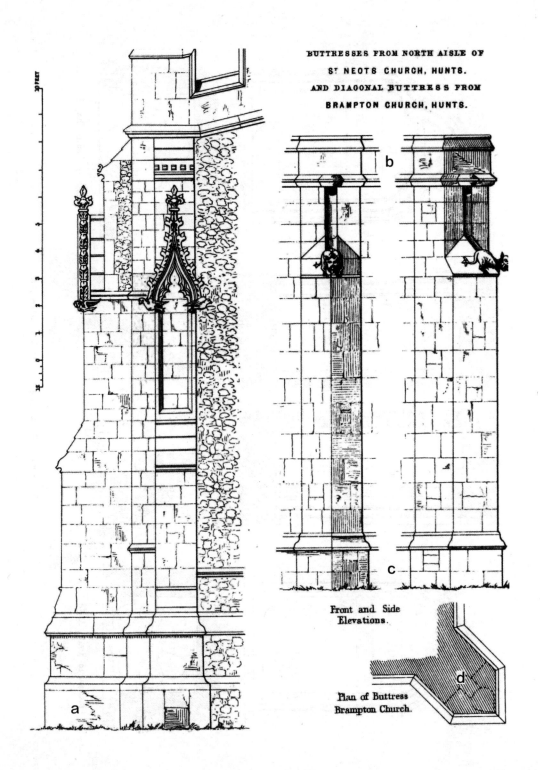

BUTTRESSES FROM NORTH AISLE OF
ST NEOTS CHURCH, HUNTS.
AND DIAGONAL BUTTRESS FROM
BRAMPTON CHURCH, HUNTS.

Front and Side
Elevations.

Plan of Buttress
Brampton Church.

a. 亨廷顿郡圣尼茨教堂北侧廊扶壁　　b. 亨廷顿郡布兰普顿教堂斜扶壁　　c. 正立面和侧立面　　d. 布兰普顿教堂扶壁平面

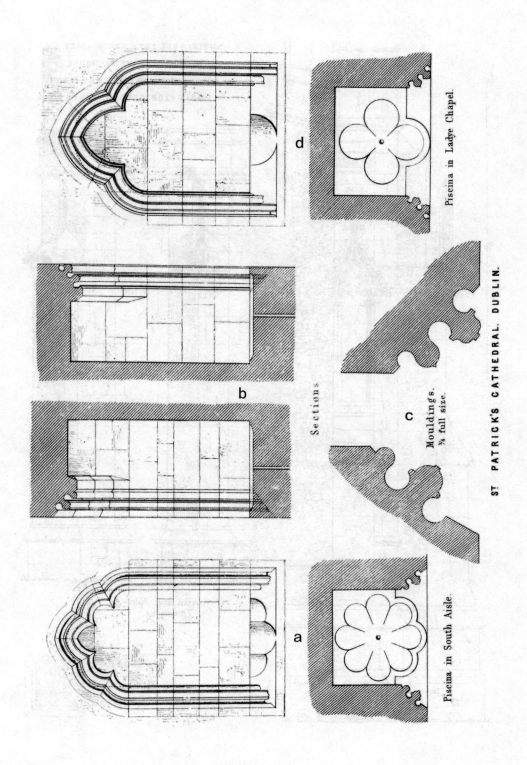

都柏林圣帕特里克教堂：a. 南走廊排水石盆　b. 剖面　c. 线脚　d. 圣母堂内排水石盆

St. Patrick's Cathedral. Dublin.
STONE CAPITALS FROM THE LADY CHAPEL.
¼ full size.

都柏林圣帕特里克教堂：a. 圣母堂内石制柱头　b. 柱础

ST. PATRICK'S CATHEDRAL, DUBLIN.
ANGLE BUTTRESSES AND CORBEL TABLE.

b
Details ¼ full size.

a

都柏林圣帕特里克教堂：a. 转角扶壁和挑檐　b. 细部

Exeter Cathedral.

HEAD OF CENTRE NICHE OF STONE REREDOS FROM THE LADY CHAPEL.

Scale 1½ inch to the foot.

埃克赛特教堂圣母堂祭坛背后石制屏风中心神龛顶盖

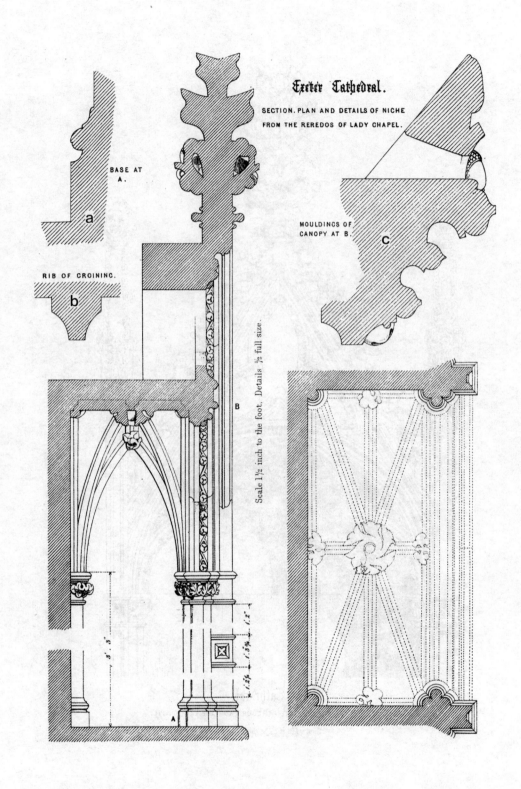

Exeter Cathedral.

SECTION. PLAN AND DETAILS OF NICHE
FROM THE REREDOS OF LADY CHAPEL.

BASE AT
A.

a

RIB OF CROINING.

b

MOULDINGS OF
CANOPY AT B.

c

Scale 1½ inch to the foot. Details ½ full size.

B

A

埃克赛特教堂圣母堂祭坛背后屏风壁龛剖面、平面及细部：a.A 处的柱础　b. 拱肋　c.B 处的顶盖线脚

2/3 full size.

2/3 full size.

Exeter Cathedral.

STONE PINNACLE FROM NICHE OF REREDOS IN THE LADY CHAPEL.

埃克赛特教堂圣母堂祭坛背后屏风壁龛上的石制小尖塔

埃克赛特教堂祭坛背后屏风上部件细部及纪念性卷叶形花饰：a. 唱诗室南侧廊纪念碑上的卷叶形花饰　b. 壁龛顶盖上的连续卷叶形花饰

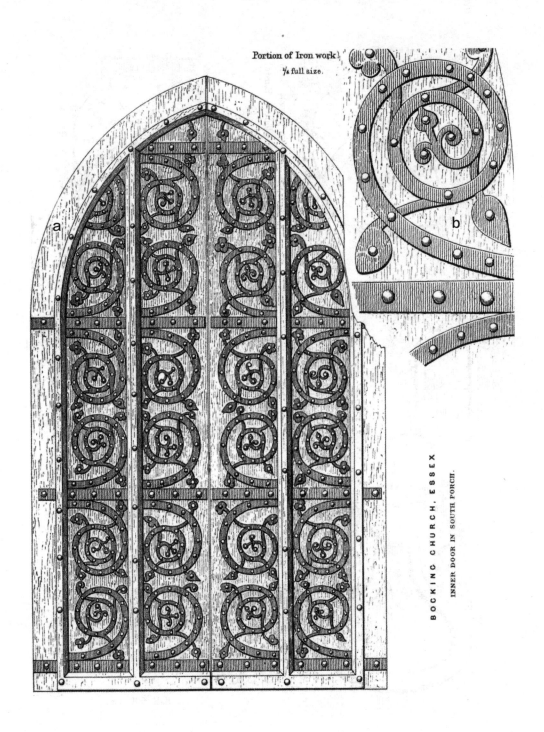

Portion of Iron work
¼ full size.

a

b

BOCKING CHURCH, ESSEX

INNER DOOR IN SOUTH PORCH.

艾塞克斯郡博金教堂：a. 南门廊内门　　b. 铁饰局部

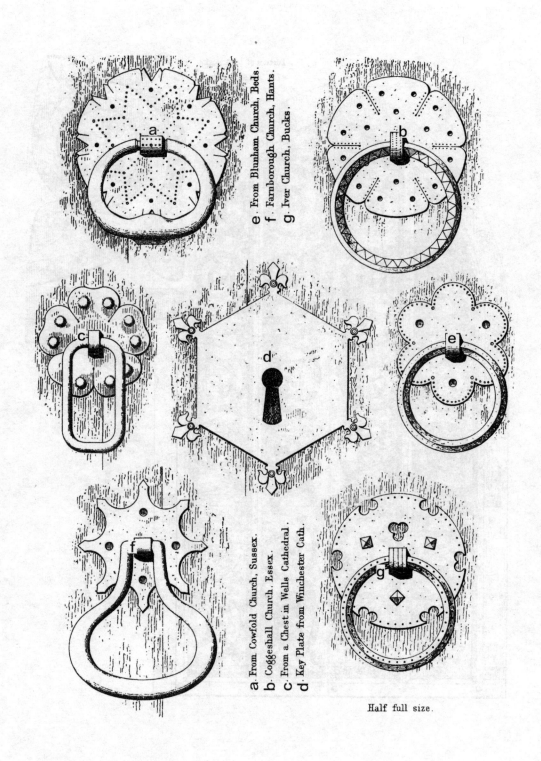

e. From Blunham Church, Beds.
f. Farnborough Church, Hants.
g. Iver Church, Bucks.

a. From Cowfold Church, Sussex.
b. Coggeshall Church, Essex.
c. From a Chest in Wells Cathedral.
d. Key Plate from Winchester Cath.

Half full size.

a. 塞萨克斯郡考福尔德教堂　b. 塞萨克斯郡科吉歇尔教堂　c. 来自威尔斯教堂一个柜子上　d. 温切斯特教堂钥匙孔板　e. 贝德弗德郡布伦海姆教堂　f. 汉普郡法恩伯勒教堂　g. 巴克斯郡艾弗教堂

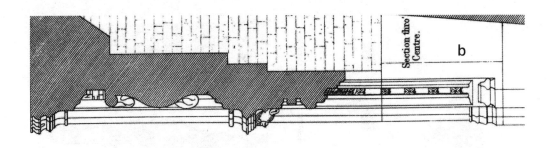

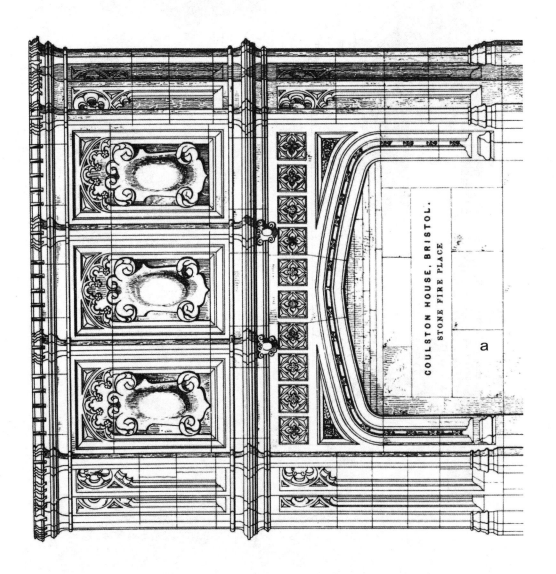

布里斯托尔科尔斯顿住所：a. 石制壁炉　b. 通过中心的剖面

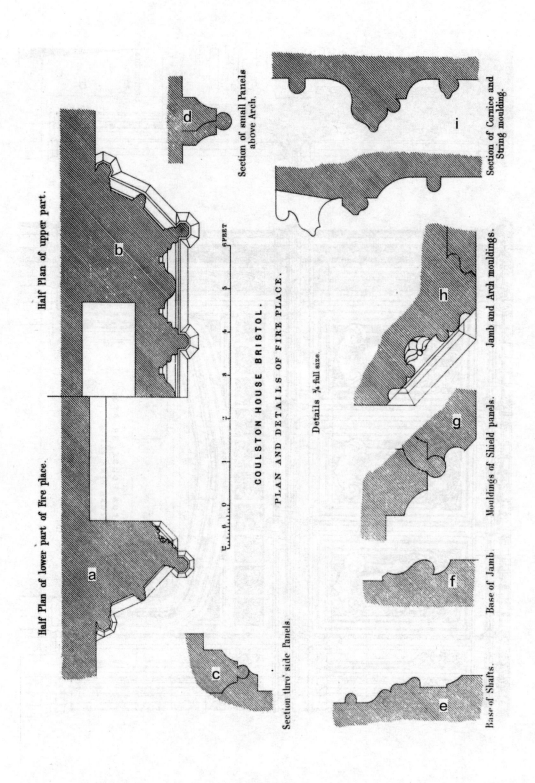

COULSTON HOUSE BRISTOL.
PLAN AND DETAILS OF FIRE PLACE.

Half Plan of upper part.
Half Plan of lower part of Fire place.
Section of small Panels above Arch.
Section of Cornice and String moulding.
Jamb and Arch mouldings.
Mouldings of Shield panels.
Base of Jamb.
Base of Shafts.
Details ¾ full size.
Section thro' side Panels.
6 FEET

布里斯托尔科尔斯顿住所壁炉平面及细部：a. 壁炉低处半平面　b. 壁炉高处半平面　c. 侧面嵌板剖面　d. 拱上小嵌板剖面　e. 柱子柱础　f. 侧壁底座　g. 防护嵌板线脚　h. 侧壁和拱线脚　i. 檐口及束带层剖面

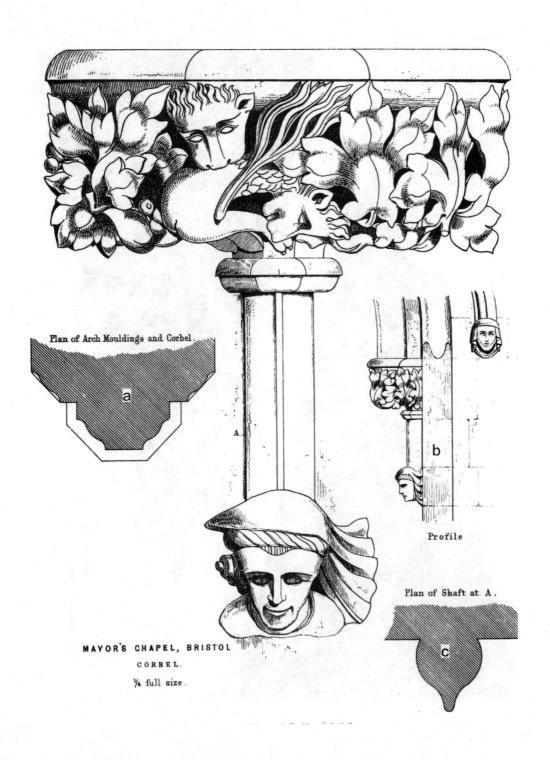

Plan of Arch Mouldings and Corbel.

a

A

b

Profile

Plan of Shaft at A.

c

MAYOR'S CHAPEL, BRISTOL
CORBEL.
¼ full size.

布里斯托尔梅厄教堂梁托：a. 拱线脚和梁托平面　b. 侧面　c.A 处柱子平面

索尔兹伯里教堂：a. 通往牧师会礼堂走廊上彩画（色彩几乎确定被修复过）　b. 穹顶平面展示了彩画的位置

索尔兹伯里教堂牧师会礼堂内铺地的瓷砖样本

诺威奇教堂唱诗室内座椅上突出托板

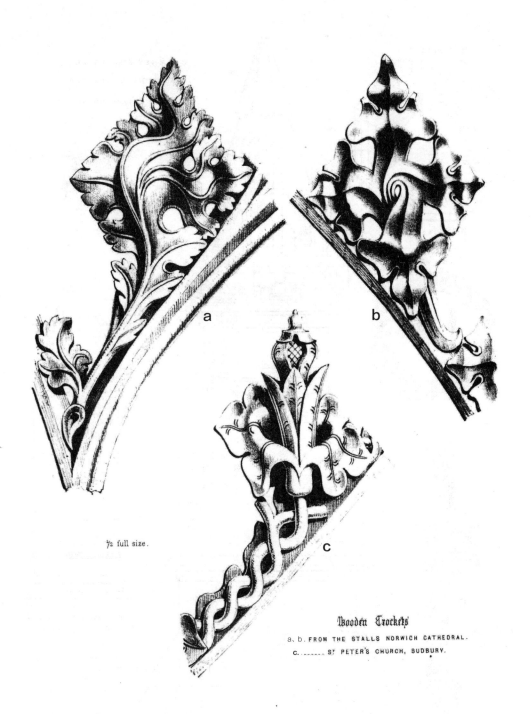

½ full size.

Wooden Crockets

a、b. FROM THE STALLS NORWICH CATHEDRAL.
c. ........ ST PETER'S CHURCH, SUDBURY.

木制卷叶形浮雕：a、b. 来自诺维奇教堂座椅　　c. 来自萨德伯里圣彼得教堂

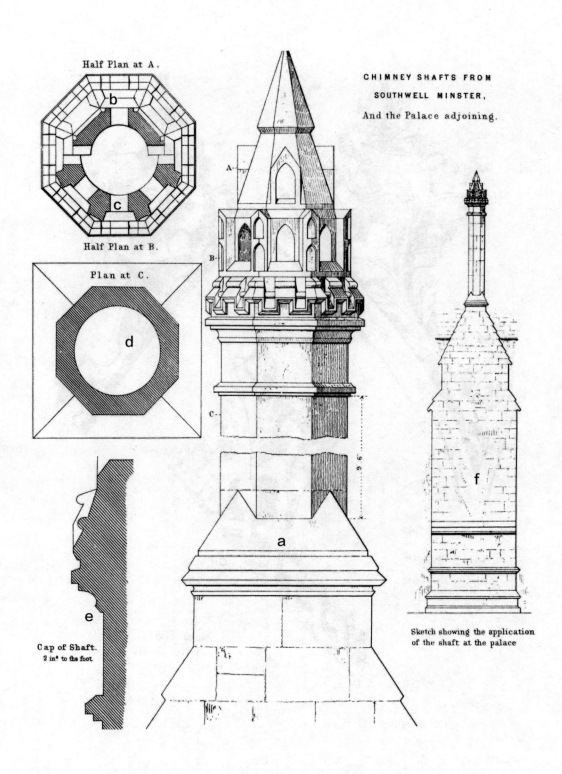

Half Plan at A.

Half Plan at B.

Plan at C.

Cap of Shaft.
2 in$^s$ to the foot

CHIMNEY SHAFTS FROM
SOUTHWELL MINSTER,
And the Palace adjoining.

Sketch showing the application
of the shaft at the palace

索思韦尔大教堂：a. 烟囱筒身　有宫殿形装饰附着　b.A 处的一半平面　c.B 处的一半平面　d.C 处的平面
e. 烟囱冒　f. 素描展示了宫殿装饰在烟囱筒上的应用

Half Plan of Window.

b

ST STEPHEN'S CHURCH,
NEAR CANTERBURY.
WINDOW IN NORTH TRANSEPT.

坎特伯雷圣史蒂芬斯教堂：a. 教堂北翼窗户    b. 窗户中部平面

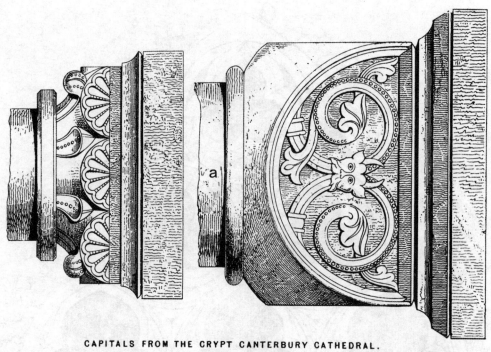

CAPITALS FROM THE CRYPT CANTERBURY CATHEDRAL.
1 AND 2 FROM CRYPT OF VESTRY.

坎特伯雷大教堂地下室柱子 a 和 b 来自于法衣室地下室

LITTLE HORMEAD CHURCH, HERTFORDSHIRE
NORTH DOOR

赫特福德郡霍米德教堂北门

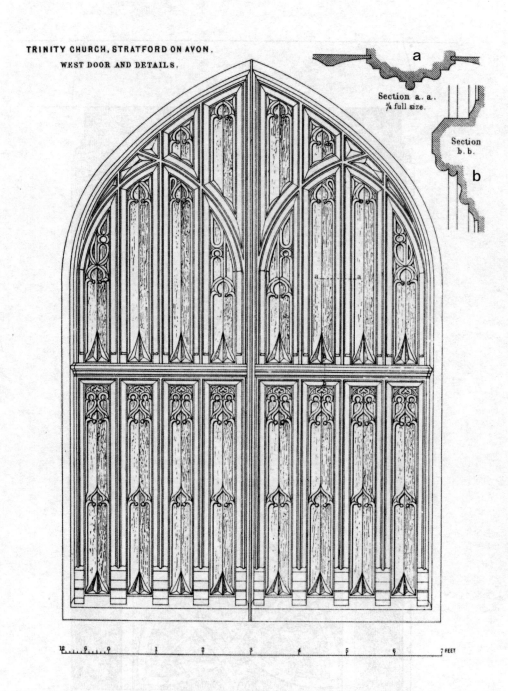

TRINITY CHURCH, STRATFORD ON AVON.
WEST DOOR AND DETAILS.

Section a. a.
¼ full size.

Section b. b.

埃文河畔斯特拉特福三一教堂西门及细部：a.a-a 剖面　b.b-b 剖面

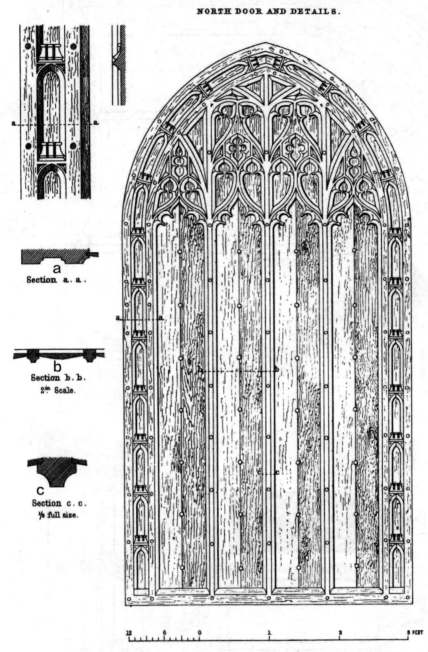

TEMPSFORD CHURCH, BEDFORDSHIRE.
NORTH DOOR AND DETAILS.

Section a. a.

Section b. b.
2ᵢₙ Scale.

Section c. c.
⅛ full size.

12    6    0    1    2    3 FEET

贝德福德郡天普斯福德教堂北门及细部：a.a-a剖面    b.b-b剖面    c.c-c剖面

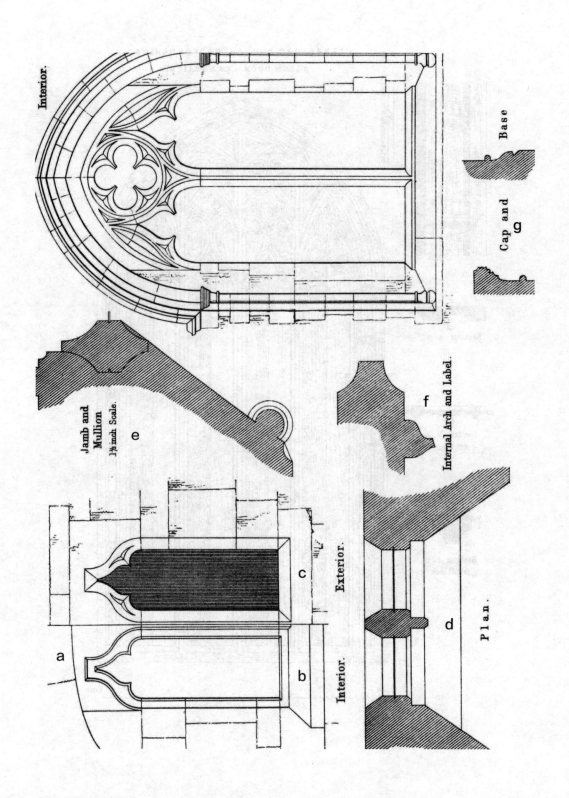

凯尼尔沃思修道院：a. 谷仓中窗户　b. 内部　c. 外部　d. 平面　e. 侧壁及竖框　f. 内部拱和披水石　g. 柱头及柱础

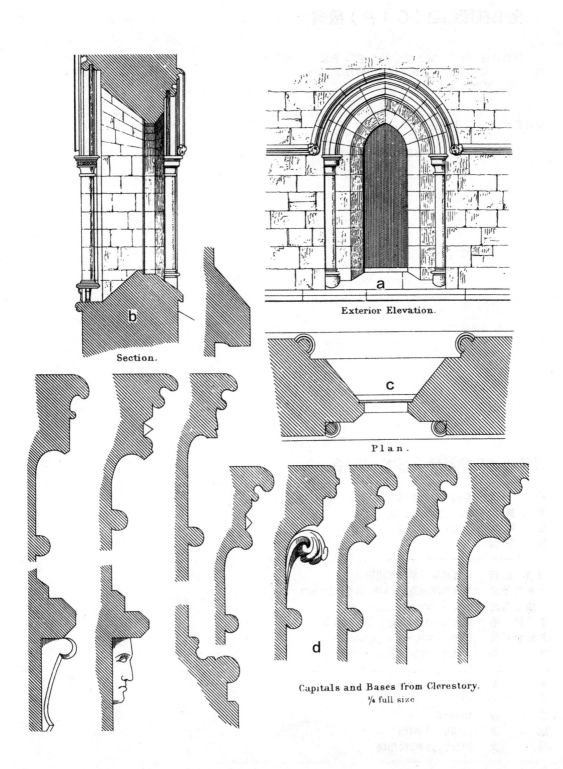

Section.

Exterior Elevation.

Plan.

Capitals and Bases from Clerestory.
¼ full size

斯卡伯勒圣玛丽教堂高窗：a. 外立面　b. 剖面　c. 平面　d. 高窗上柱头和柱础

# 图书在版编目（CIP）数据

哥特建筑与雕塑装饰艺术．第2卷／曹峻川，甄影博
编．－－南京：江苏凤凰科学技术出版社，2018.1
ISBN 978-7-5537-8756-5

Ⅰ．①哥… Ⅱ．①曹…②甄… Ⅲ．①哥特式建筑－
建筑艺术 Ⅳ．① TU-098.2

中国版本图书馆 CIP 数据核字 (2017) 第 292846 号

## 哥特建筑与雕塑装饰艺术　第2卷

| | | |
|---|---|---|
| 编　　　译 | 曹峻川　甄影博 | |
| 项 目 策 划 | 凤凰空间／郑亚男 | |
| 责 任 编 辑 | 刘屹立　赵　研 | |
| 特 约 编 辑 | 苑　圆 | |

| | |
|---|---|
| 出 版 发 行 | 江苏凤凰科学技术出版社 |
| 出版社地址 | 南京市湖南路1号A楼 邮编：210009 |
| 出版社网址 | http://www.pspress.cn |
| 总 经 销 | 天津凤凰空间文化传媒有限公司 |
| 总经销网址 | http://www.ifengspace.cn |
| 印　　　刷 | 北京建宏印刷有限公司 |

| | |
|---|---|
| 开　　　本 | 710 mm×1000 mm　1／8 |
| 印　　　张 | 40 |
| 字　　　数 | 160 000 |
| 版　　　次 | 2018年1月第1版 |
| 印　　　次 | 2023年3月第2次印刷 |

| | |
|---|---|
| 标 准 书 号 | ISBN 978-7-5537-8756-5 |
| 定　　　价 | 188.00元 |

图书如有印装质量问题，可随时向销售部调换（电话：022-87893668）。